American Radio Club's

Amateur Radio Station Logbook

Name:

_____

Call:

_____

Copyright © 2022 by American Radio Club

All rights reserved. No part of this publication may be reproduced, distributed or transmitted in any form or by any means, including photocopying, recording, or other electronic or mechanical methods, without the prior written permission of the publisher, except in the case of brief quotations embodied in critical reviews and certain other noncommercial uses permitted by copyright law. For permission requests, write to the publisher, addressed "Attention: Permissions Coordinator," at the address below.

1309 Coffeen Ave Ste 1956
Sheridan WY 82801

Or visit: AmericanRadioClub.com

Ordering Information:

Special discounts are available on quantity purchases by corporations, associations and others. For details, contact the publisher at the address below.

Email: support@AmericanRadioClub.com

Printed in the United States of America

ISBN: 9798746032270

Imprint: Independently published

 AMERICAN RADIO CLUB

Operators Creed

Licensed amateur radio operators play an important role in our nation and world. They :

Operate on the air legally and follow Federal Communications Commission rules and side agreements that allow others to use the airwaves or furtherance of the hobby.

Use their radio equipment and stations in a way that benefits themselves as a hobby but also their community and nation, and even worlds. Amateur radio is valuable resource that always must be put to good use for the benefit of all.

Utilize spectrum efficiently and properly so that amateur radio has a presence throughout the spectrum to ensure frequencies are available for generations of new amateurs to come.

Embrace emerging technologies and digital standards and platforms so that they are advancing the art of amateur radio by being on the air.

Respect fellow amateurs on the air and ensure that amateur radio frequencies are a welcoming place for all to gather, no matter their background.

Prepare others to learn the science and art of amateur radio so that we always are increasing our ranks to ensure that our spectrum is utilized as much as possible.

Use the hobby to learn not only more about science and technology, but also our family and neighbors.

About this logbook

While the Federal Communications Commission hasn't required U.S. amateur radio operators to maintain logbooks for decades, many still prefer to keep a record of their on-air contacts as a matter of record and for QSL purposes.

American Radio Club has made it easy to use this logbook with minimal information to find your logging activity useful. Most amateurs log in UTC (Coordinated Universal Time, also known as Greenwich Mean Time) for date and time.

There also is room for you to log the call sign of the station you have contacted, as well as the frequency or band you are operating on.

The notes section can be used to make notations as to the other station's location, RST signal reports exchanged over the air, modes, power used, and even whether or not you sent or received a QSL card.

We look forward to hearing you on the air soon.

AMERICAN RADIO CLUB

Date	Time (UTC)	Frequency or Band	Mode	Power	Station worked	Report sent	Report rec'd	QTH and Name	QSL sent	QSL rec'd
									☐	☐
									☐	☐
									☐	☐
									☐	☐
									☐	☐
									☐	☐
									☐	☐
									☐	☐
									☐	☐
									☐	☐

AMERICAN RADIO CLUB
Notes

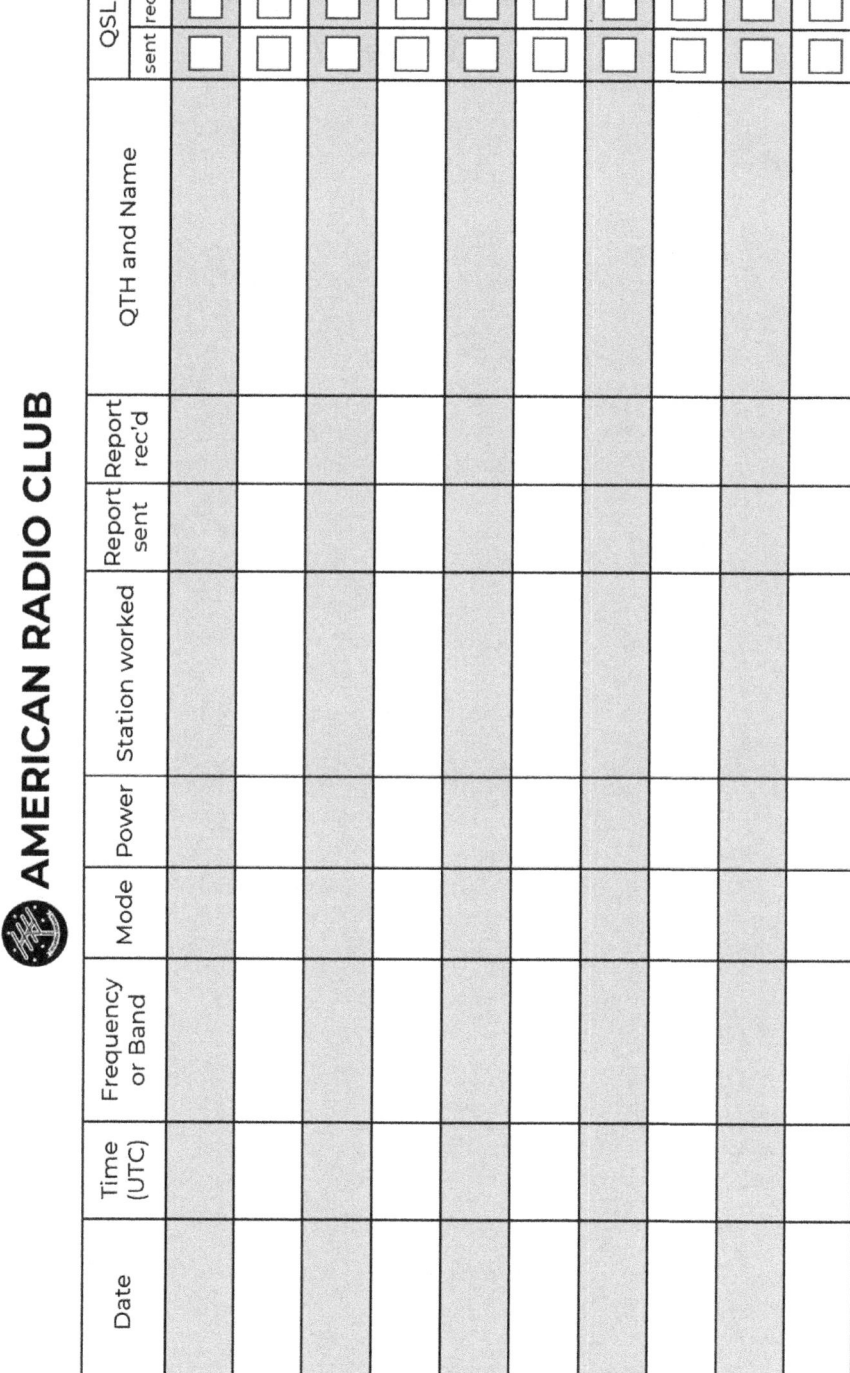

AMERICAN RADIO CLUB
Notes

AMERICAN RADIO CLUB

Date	Time (UTC)	Frequency or Band	Mode	Power	Station worked	Report sent	Report rec'd	QTH and Name	QSL sent	QSL rec'd
									☐	☐
									☐	☐
									☐	☐
									☐	☐
									☐	☐
									☐	☐
									☐	☐
									☐	☐
									☐	☐
									☐	☐

AMERICAN RADIO CLUB
Notes

AMERICAN RADIO CLUB

Date	Time (UTC)	Frequency or Band	Mode	Power	Station worked	Report sent	Report rec'd	QTH and Name	QSL sent	QSL rec'd
									☐	☐
									☐	☐
									☐	☐
									☐	☐
									☐	☐
									☐	☐
									☐	☐
									☐	☐
									☐	☐
									☐	☐

AMERICAN RADIO CLUB
Notes

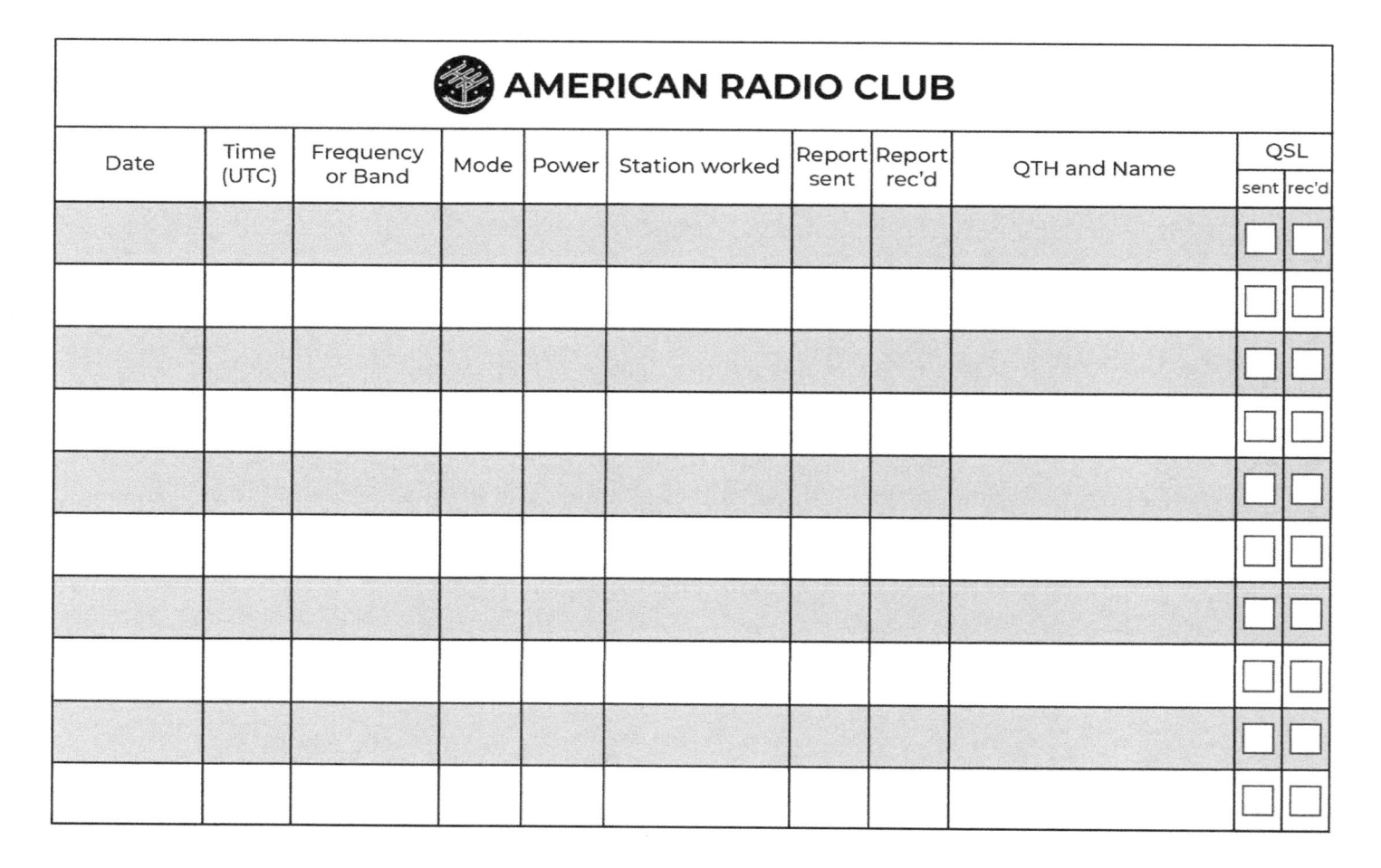

AMERICAN RADIO CLUB

Date	Time (UTC)	Frequency or Band	Mode	Power	Station worked	Report sent	Report rec'd	QTH and Name	QSL sent	QSL rec'd
									☐	☐
									☐	☐
									☐	☐
									☐	☐
									☐	☐
									☐	☐
									☐	☐
									☐	☐
									☐	☐
									☐	☐

AMERICAN RADIO CLUB
Notes

AMERICAN RADIO CLUB

Date	Time (UTC)	Frequency or Band	Mode	Power	Station worked	Report sent	Report rec'd	QTH and Name	QSL sent	QSL rec'd
									☐	☐
									☐	☐
									☐	☐
									☐	☐
									☐	☐
									☐	☐
									☐	☐
									☐	☐
									☐	☐
									☐	☐

AMERICAN RADIO CLUB
Notes

… AMERICAN RADIO CLUB

Date	Time (UTC)	Frequency or Band	Mode	Power	Station worked	Report sent	Report rec'd	QTH and Name	QSL sent	QSL rec'd
									☐	☐
									☐	☐
									☐	☐
									☐	☐
									☐	☐
									☐	☐
									☐	☐
									☐	☐
									☐	☐
									☐	☐

AMERICAN RADIO CLUB
Notes

AMERICAN RADIO CLUB

Date	Time (UTC)	Frequency or Band	Mode	Power	Station worked	Report sent	Report rec'd	QTH and Name	QSL sent	QSL rec'd
									☐	☐
									☐	☐
									☐	☐
									☐	☐
									☐	☐
									☐	☐
									☐	☐
									☐	☐
									☐	☐
									☐	☐

AMERICAN RADIO CLUB
Notes

AMERICAN RADIO CLUB

Date	Time (UTC)	Frequency or Band	Mode	Power	Station worked	Report sent	Report rec'd	QTH and Name	QSL sent	QSL rec'd
									☐	☐
									☐	☐
									☐	☐
									☐	☐
									☐	☐
									☐	☐
									☐	☐
									☐	☐
									☐	☐
									☐	☐

AMERICAN RADIO CLUB
Notes

AMERICAN RADIO CLUB

Date	Time (UTC)	Frequency or Band	Mode	Power	Station worked	Report sent	Report rec'd	QTH and Name	QSL sent	QSL rec'd
									☐	☐
									☐	☐
									☐	☐
									☐	☐
									☐	☐
									☐	☐
									☐	☐
									☐	☐
									☐	☐
									☐	☐

AMERICAN RADIO CLUB
Notes

AMERICAN RADIO CLUB

Date	Time (UTC)	Frequency or Band	Mode	Power	Station worked	Report sent	Report rec'd	QTH and Name	QSL sent	QSL rec'd
									☐	☐
									☐	☐
									☐	☐
									☐	☐
									☐	☐
									☐	☐
									☐	☐
									☐	☐
									☐	☐
									☐	☐

AMERICAN RADIO CLUB
Notes

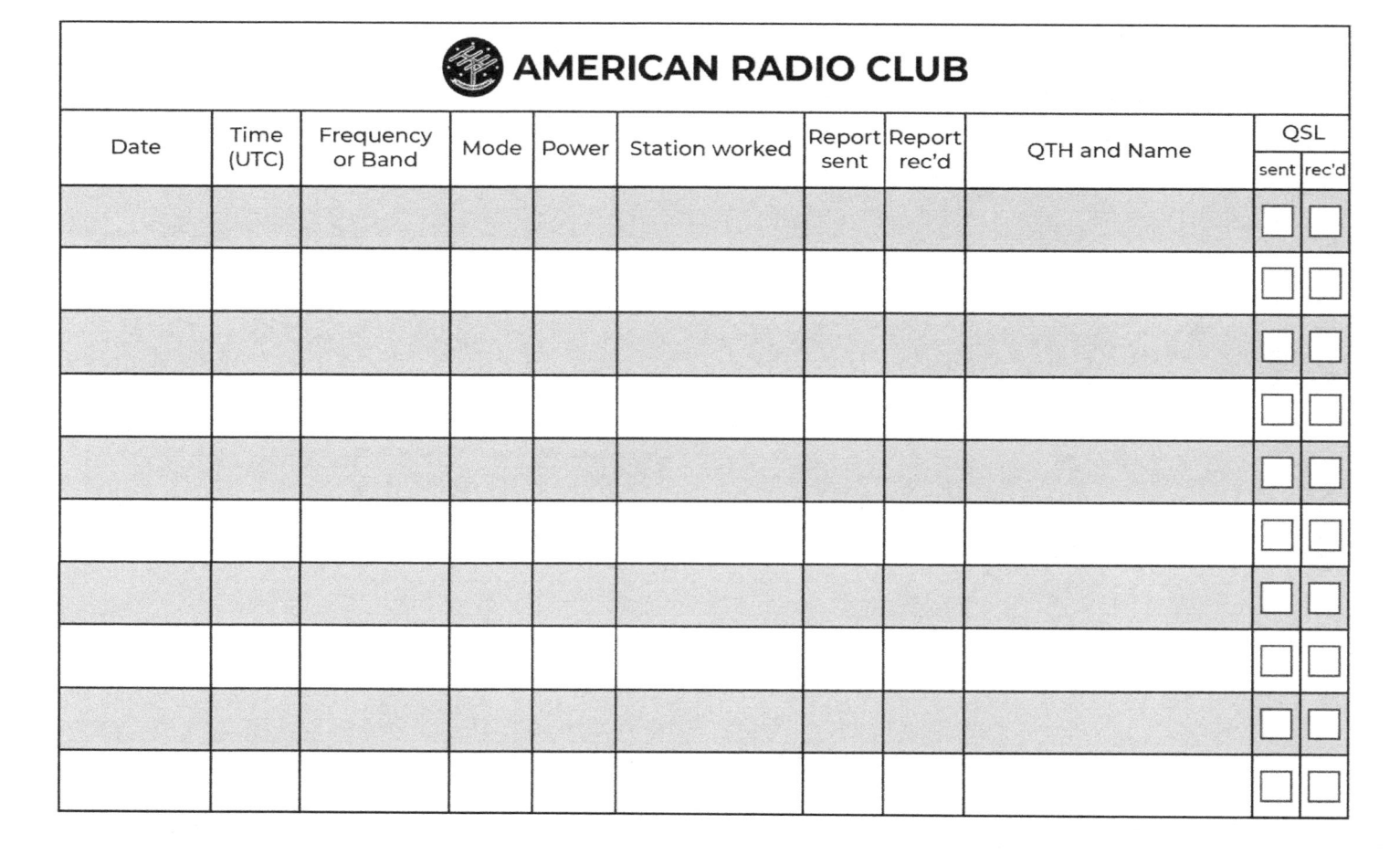

AMERICAN RADIO CLUB

Date	Time (UTC)	Frequency or Band	Mode	Power	Station worked	Report sent	Report rec'd	QTH and Name	QSL sent	QSL rec'd
									☐	☐
									☐	☐
									☐	☐
									☐	☐
									☐	☐
									☐	☐
									☐	☐
									☐	☐
									☐	☐

AMERICAN RADIO CLUB
Notes

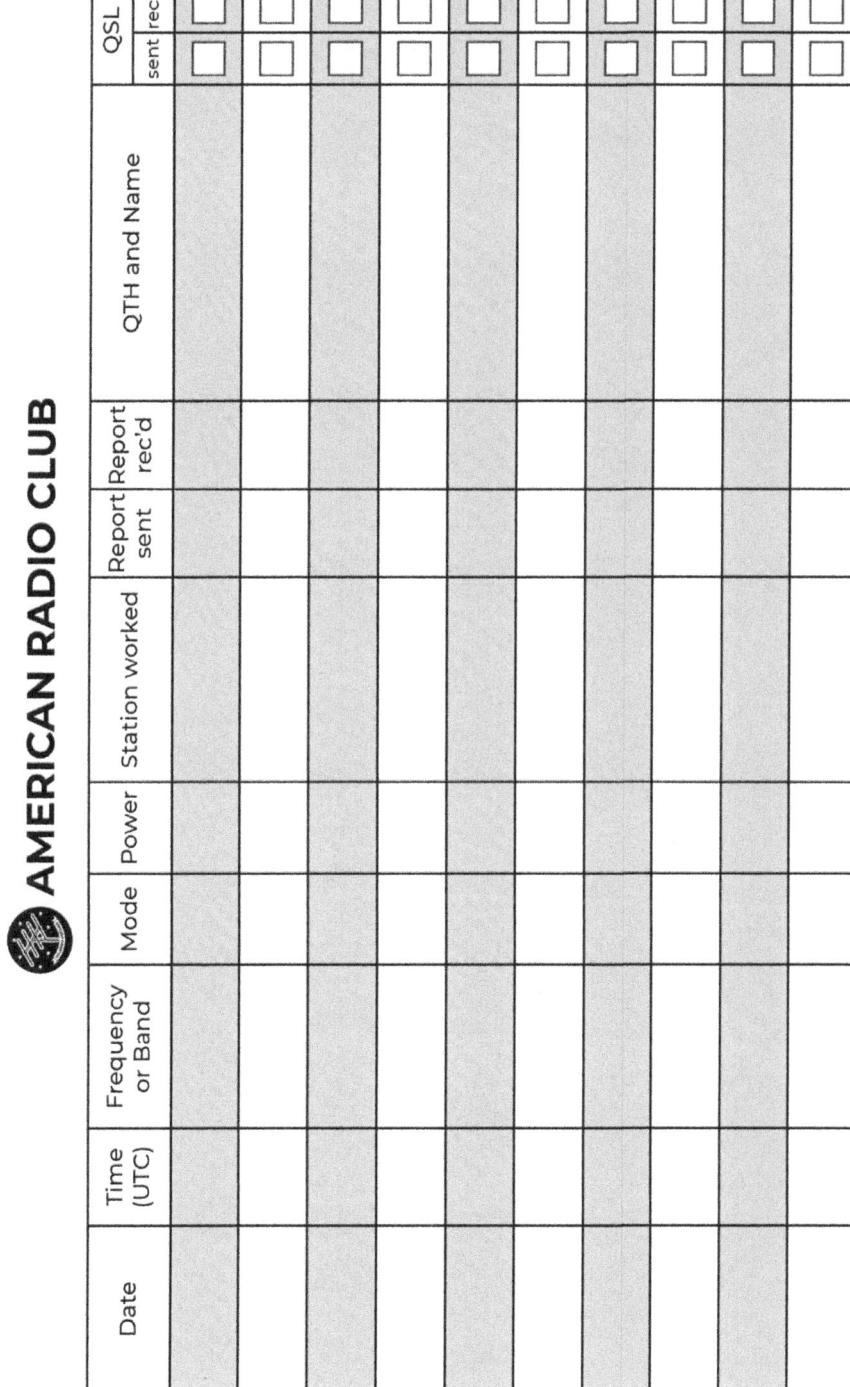

AMERICAN RADIO CLUB
Notes

AMERICAN RADIO CLUB

Date	Time (UTC)	Frequency or Band	Mode	Power	Station worked	Report sent	Report rec'd	QTH and Name	QSL sent	QSL rec'd
									☐	☐
									☐	☐
									☐	☐
									☐	☐
									☐	☐
									☐	☐
									☐	☐
									☐	☐
									☐	☐
									☐	☐

AMERICAN RADIO CLUB
Notes

AMERICAN RADIO CLUB

Date	Time (UTC)	Frequency or Band	Mode	Power	Station worked	Report sent	Report rec'd	QTH and Name	QSL sent	QSL rec'd
									☐	☐
									☐	☐
									☐	☐
									☐	☐
									☐	☐
									☐	☐
									☐	☐
									☐	☐
									☐	☐
									☐	☐

AMERICAN RADIO CLUB
Notes

AMERICAN RADIO CLUB

Date	Time (UTC)	Frequency or Band	Mode	Power	Station worked	Report sent	Report rec'd	QTH and Name	QSL sent	QSL rec'd
									☐	☐
									☐	☐
									☐	☐
									☐	☐
									☐	☐
									☐	☐
									☐	☐
									☐	☐
									☐	☐
									☐	☐

AMERICAN RADIO CLUB
Notes

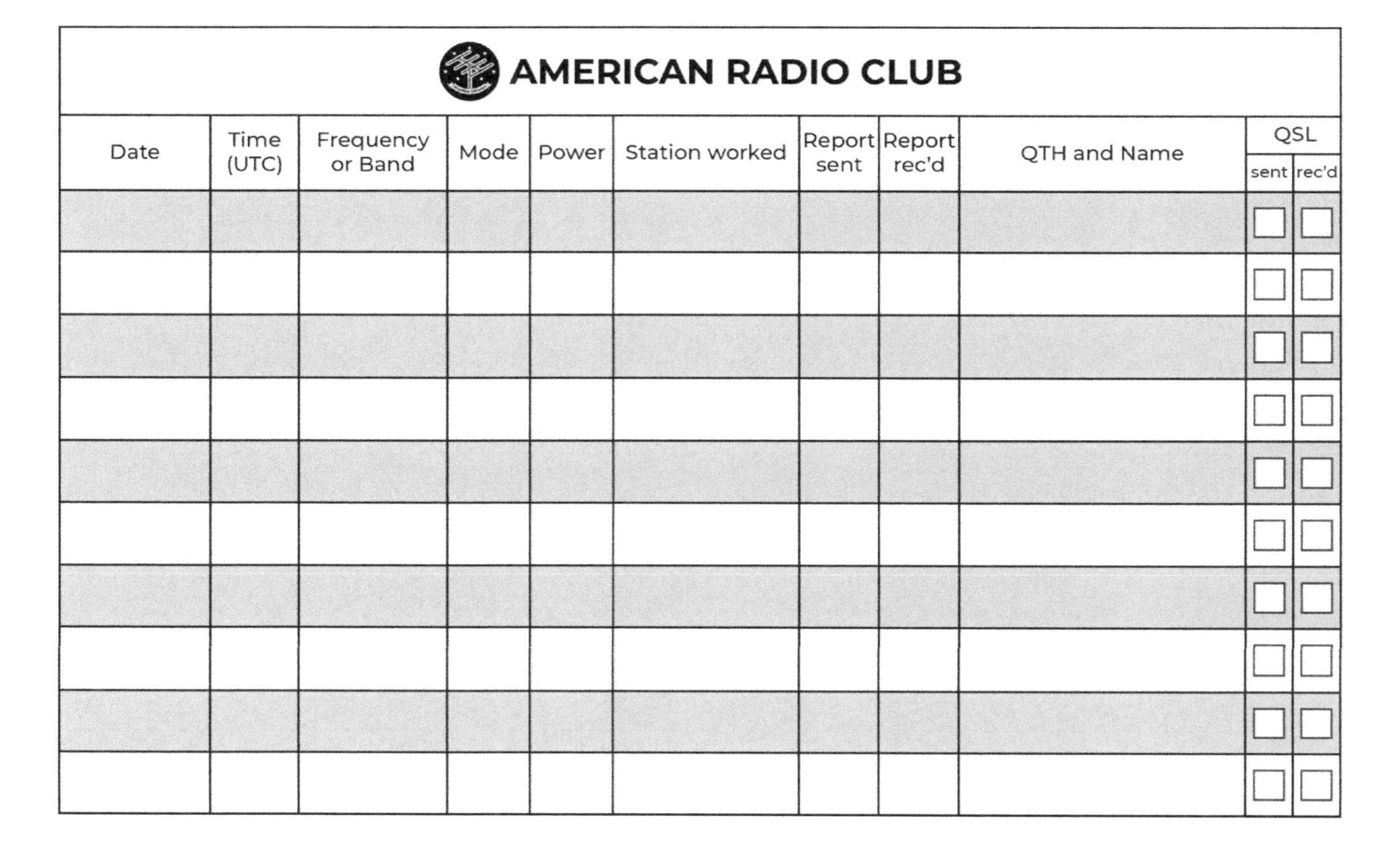

AMERICAN RADIO CLUB

Date	Time (UTC)	Frequency or Band	Mode	Power	Station worked	Report sent	Report rec'd	QTH and Name	QSL sent	QSL rec'd
									☐	☐
									☐	☐
									☐	☐
									☐	☐
									☐	☐
									☐	☐
									☐	☐
									☐	☐
									☐	☐
									☐	☐

AMERICAN RADIO CLUB
Notes

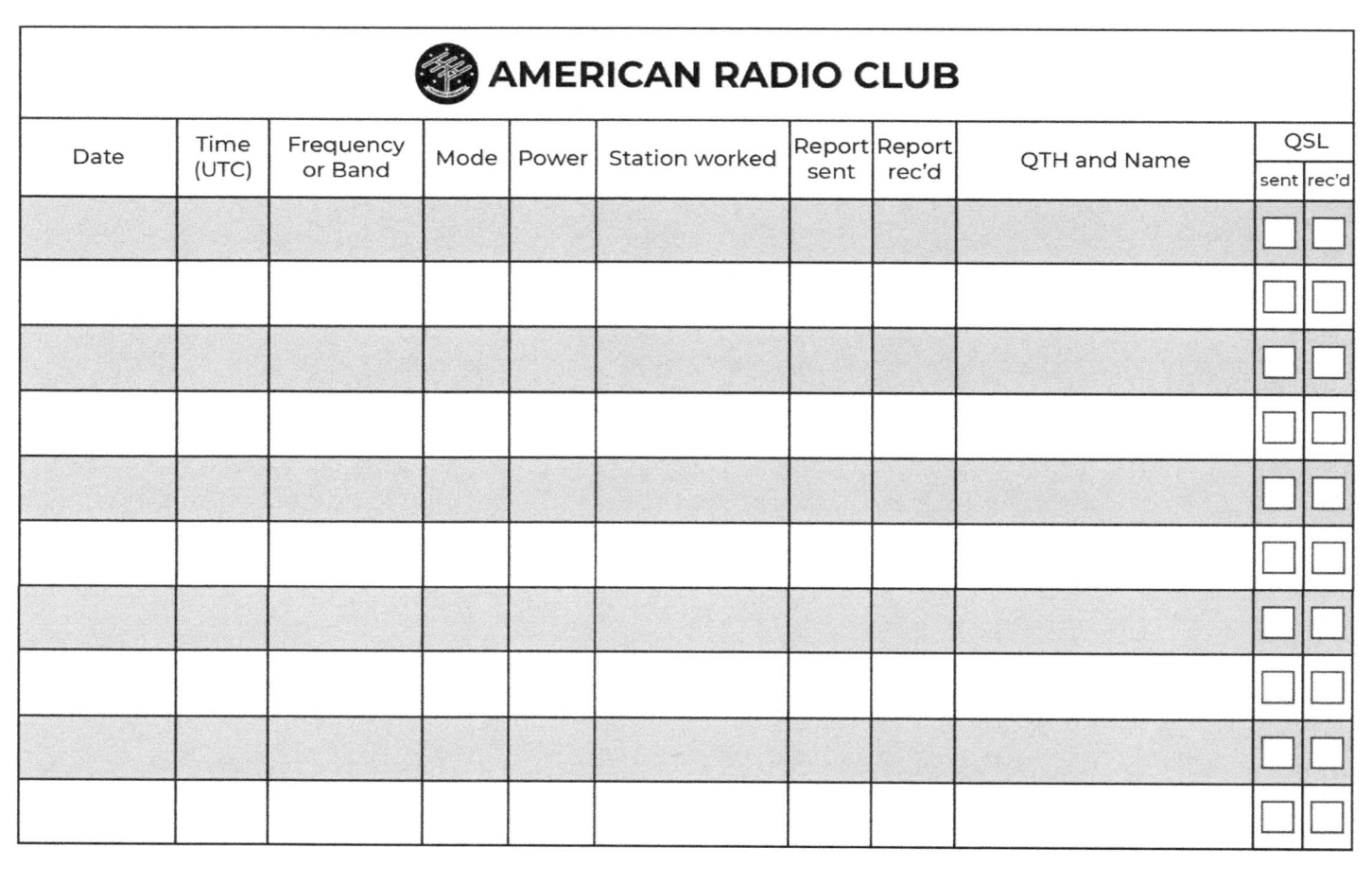

AMERICAN RADIO CLUB

Date	Time (UTC)	Frequency or Band	Mode	Power	Station worked	Report sent	Report rec'd	QTH and Name	QSL sent	QSL rec'd
									☐	☐
									☐	☐
									☐	☐
									☐	☐
									☐	☐
									☐	☐
									☐	☐
									☐	☐
									☐	☐
									☐	☐

AMERICAN RADIO CLUB
Notes

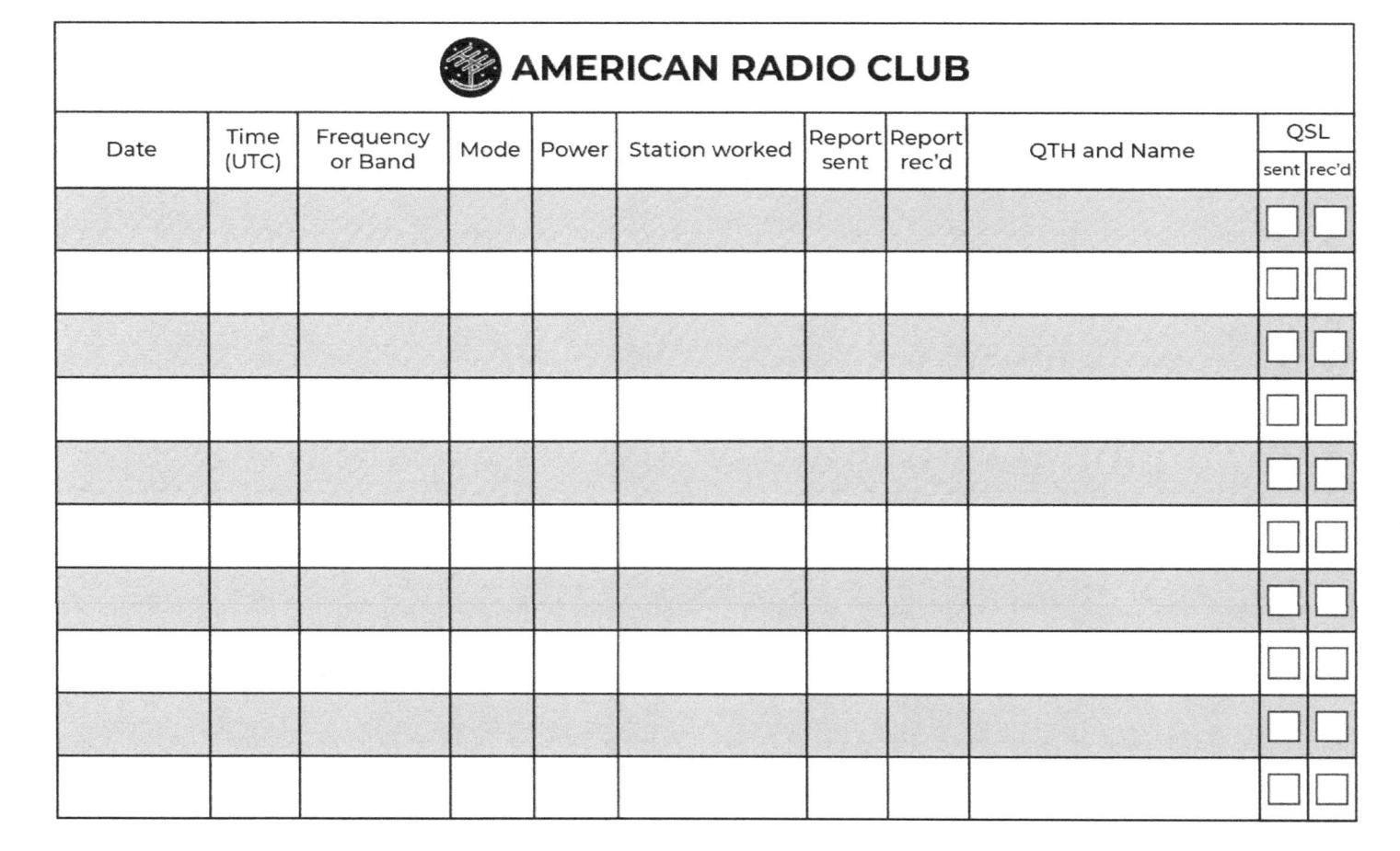

… AMERICAN RADIO CLUB

Date	Time (UTC)	Frequency or Band	Mode	Power	Station worked	Report sent	Report rec'd	QTH and Name	QSL sent	QSL rec'd
									☐	☐
									☐	☐
									☐	☐
									☐	☐
									☐	☐
									☐	☐
									☐	☐
									☐	☐
									☐	☐
									☐	☐

AMERICAN RADIO CLUB
Notes

AMERICAN RADIO CLUB

Date	Time (UTC)	Frequency or Band	Mode	Power	Station worked	Report sent	Report rec'd	QTH and Name	QSL sent	QSL rec'd
									☐	☐
									☐	☐
									☐	☐
									☐	☐
									☐	☐
									☐	☐
									☐	☐
									☐	☐
									☐	☐
									☐	☐

AMERICAN RADIO CLUB
Notes

AMERICAN RADIO CLUB

Date	Time (UTC)	Frequency or Band	Mode	Power	Station worked	Report sent	Report rec'd	QTH and Name	QSL sent	QSL rec'd
									☐	☐
									☐	☐
									☐	☐
									☐	☐
									☐	☐
									☐	☐
									☐	☐
									☐	☐
									☐	☐
									☐	☐

AMERICAN RADIO CLUB
Notes

AMERICAN RADIO CLUB

Date	Time (UTC)	Frequency or Band	Mode	Power	Station worked	Report sent	Report rec'd	QTH and Name	QSL sent	QSL rec'd
									☐	☐
									☐	☐
									☐	☐
									☐	☐
									☐	☐
									☐	☐
									☐	☐
									☐	☐
									☐	☐
									☐	☐

AMERICAN RADIO CLUB
Notes

AMERICAN RADIO CLUB

Date	Time (UTC)	Frequency or Band	Mode	Power	Station worked	Report sent	Report rec'd	QTH and Name	QSL sent	QSL rec'd
									☐	☐
									☐	☐
									☐	☐
									☐	☐
									☐	☐
									☐	☐
									☐	☐
									☐	☐
									☐	☐
									☐	☐

AMERICAN RADIO CLUB
Notes

AMERICAN RADIO CLUB

Date	Time (UTC)	Frequency or Band	Mode	Power	Station worked	Report sent	Report rec'd	QTH and Name	QSL sent	QSL rec'd
									☐	☐
									☐	☐
									☐	☐
									☐	☐
									☐	☐
									☐	☐
									☐	☐
									☐	☐
									☐	☐
									☐	☐

AMERICAN RADIO CLUB
Notes

AMERICAN RADIO CLUB

Date	Time (UTC)	Frequency or Band	Mode	Power	Station worked	Report sent	Report rec'd	QTH and Name	QSL sent	QSL rec'd
									☐	☐
									☐	☐
									☐	☐
									☐	☐
									☐	☐
									☐	☐
									☐	☐
									☐	☐
									☐	☐
									☐	☐

AMERICAN RADIO CLUB
Notes

AMERICAN RADIO CLUB

Date	Time (UTC)	Frequency or Band	Mode	Power	Station worked	Report sent	Report rec'd	QTH and Name	QSL sent	QSL rec'd
									☐	☐
									☐	☐
									☐	☐
									☐	☐
									☐	☐
									☐	☐
									☐	☐
									☐	☐
									☐	☐
									☐	☐

AMERICAN RADIO CLUB
Notes

AMERICAN RADIO CLUB

Date	Time (UTC)	Frequency or Band	Mode	Power	Station worked	Report sent	Report rec'd	QTH and Name	QSL sent	QSL rec'd
									☐	☐
									☐	☐
									☐	☐
									☐	☐
									☐	☐
									☐	☐
									☐	☐
									☐	☐
									☐	☐
									☐	☐

AMERICAN RADIO CLUB
Notes

AMERICAN RADIO CLUB

Date	Time (UTC)	Frequency or Band	Mode	Power	Station worked	Report sent	Report rec'd	QTH and Name	QSL sent	QSL rec'd
									☐	☐
									☐	☐
									☐	☐
									☐	☐
									☐	☐
									☐	☐
									☐	☐
									☐	☐
									☐	☐

AMERICAN RADIO CLUB
Notes

AMERICAN RADIO CLUB

Date	Time (UTC)	Frequency or Band	Mode	Power	Station worked	Report sent	Report rec'd	QTH and Name	QSL sent	QSL rec'd
									☐	☐
									☐	☐
									☐	☐
									☐	☐
									☐	☐
									☐	☐
									☐	☐
									☐	☐
									☐	☐
									☐	☐

AMERICAN RADIO CLUB
Notes

AMERICAN RADIO CLUB

Date	Time (UTC)	Frequency or Band	Mode	Power	Station worked	Report sent	Report rec'd	QTH and Name	QSL sent	QSL rec'd
									☐	☐
									☐	☐
									☐	☐
									☐	☐
									☐	☐
									☐	☐
									☐	☐
									☐	☐
									☐	☐
									☐	☐

AMERICAN RADIO CLUB
Notes

AMERICAN RADIO CLUB

Date	Time (UTC)	Frequency or Band	Mode	Power	Station worked	Report sent	Report rec'd	QTH and Name	QSL sent	QSL rec'd
									☐	☐
									☐	☐
									☐	☐
									☐	☐
									☐	☐
									☐	☐
									☐	☐
									☐	☐
									☐	☐
									☐	☐

AMERICAN RADIO CLUB
Notes

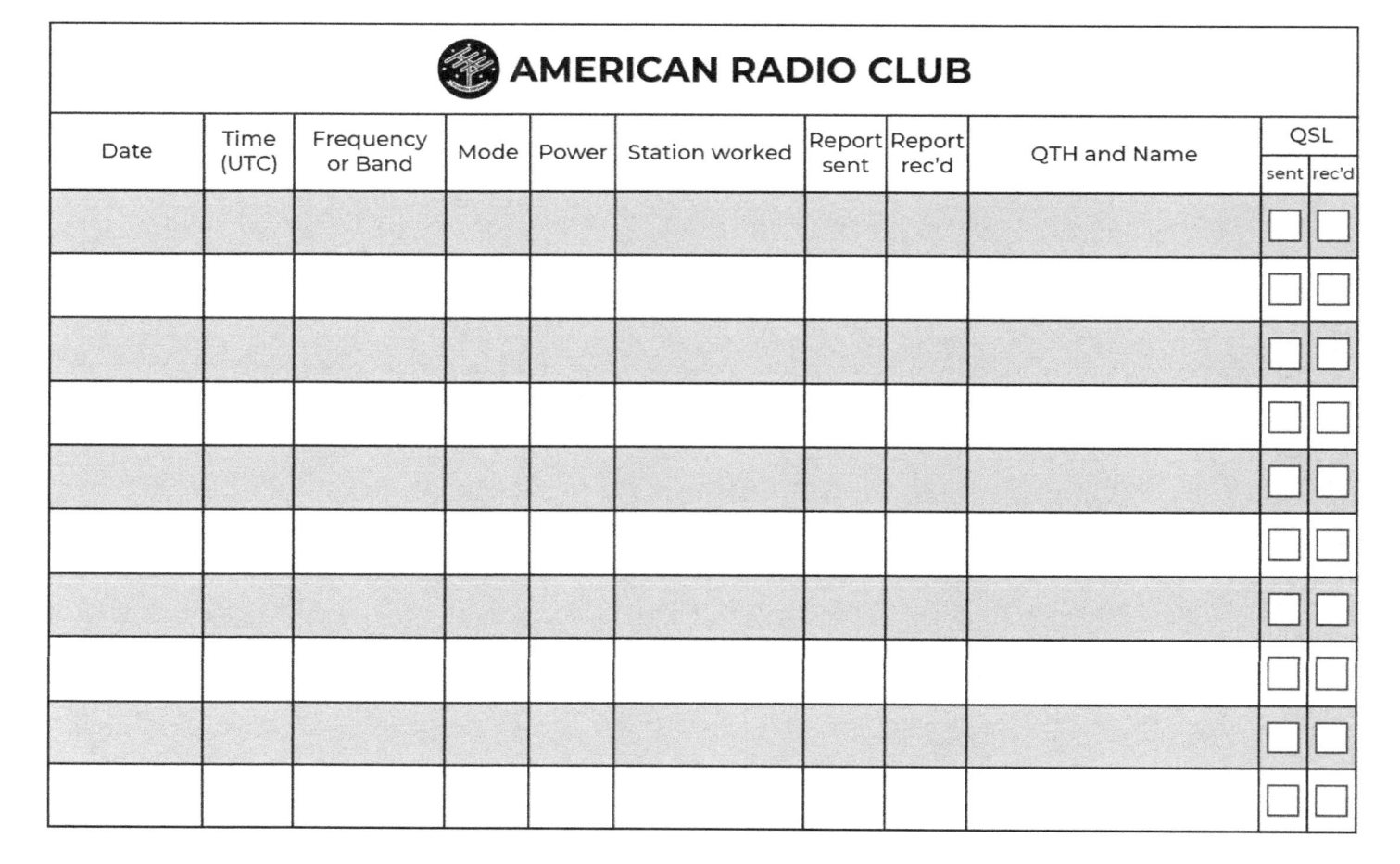

AMERICAN RADIO CLUB

Date	Time (UTC)	Frequency or Band	Mode	Power	Station worked	Report sent	Report rec'd	QTH and Name	QSL sent	QSL rec'd
									☐	☐
									☐	☐
									☐	☐
									☐	☐
									☐	☐
									☐	☐
									☐	☐
									☐	☐
									☐	☐
									☐	☐

AMERICAN RADIO CLUB
Notes

AMERICAN RADIO CLUB

Date	Time (UTC)	Frequency or Band	Mode	Power	Station worked	Report sent	Report rec'd	QTH and Name	QSL sent	QSL rec'd
									☐	☐
									☐	☐
									☐	☐
									☐	☐
									☐	☐
									☐	☐
									☐	☐
									☐	☐
									☐	☐
									☐	☐

AMERICAN RADIO CLUB
Notes

AMERICAN RADIO CLUB

Date	Time (UTC)	Frequency or Band	Mode	Power	Station worked	Report sent	Report rec'd	QTH and Name	QSL sent	QSL rec'd
									☐	☐
									☐	☐
									☐	☐
									☐	☐
									☐	☐
									☐	☐
									☐	☐
									☐	☐
									☐	☐
									☐	☐

AMERICAN RADIO CLUB
Notes

AMERICAN RADIO CLUB

Date	Time (UTC)	Frequency or Band	Mode	Power	Station worked	Report sent	Report rec'd	QTH and Name	QSL sent	QSL rec'd
									☐	☐
									☐	☐
									☐	☐
									☐	☐
									☐	☐
									☐	☐
									☐	☐
									☐	☐
									☐	☐
									☐	☐

AMERICAN RADIO CLUB
Notes

AMERICAN RADIO CLUB

Date	Time (UTC)	Frequency or Band	Mode	Power	Station worked	Report sent	Report rec'd	QTH and Name	QSL sent	QSL rec'd
									☐	☐
									☐	☐
									☐	☐
									☐	☐
									☐	☐
									☐	☐
									☐	☐
									☐	☐
									☐	☐
									☐	☐

… AMERICAN RADIO CLUB
Notes

AMERICAN RADIO CLUB

Date	Time (UTC)	Frequency or Band	Mode	Power	Station worked	Report sent	Report rec'd	QTH and Name	QSL sent	QSL rec'd
									☐	☐
									☐	☐
									☐	☐
									☐	☐
									☐	☐
									☐	☐
									☐	☐
									☐	☐
									☐	☐
									☐	☐

AMERICAN RADIO CLUB
Notes

AMERICAN RADIO CLUB

Date	Time (UTC)	Frequency or Band	Mode	Power	Station worked	Report sent	Report rec'd	QTH and Name	QSL sent	QSL rec'd
									☐	☐
									☐	☐
									☐	☐
									☐	☐
									☐	☐
									☐	☐
									☐	☐
									☐	☐
									☐	☐
									☐	☐

AMERICAN RADIO CLUB
Notes

AMERICAN RADIO CLUB

Date	Time (UTC)	Frequency or Band	Mode	Power	Station worked	Report sent	Report rec'd	QTH and Name	QSL sent	QSL rec'd
									☐	☐
									☐	☐
									☐	☐
									☐	☐
									☐	☐
									☐	☐
									☐	☐
									☐	☐
									☐	☐
									☐	☐

AMERICAN RADIO CLUB
Notes

AMERICAN RADIO CLUB

Date	Time (UTC)	Frequency or Band	Mode	Power	Station worked	Report sent	Report rec'd	QTH and Name	QSL sent	QSL rec'd
									☐	☐
									☐	☐
									☐	☐
									☐	☐
									☐	☐
									☐	☐
									☐	☐
									☐	☐
									☐	☐
									☐	☐

AMERICAN RADIO CLUB
Notes

AMERICAN RADIO CLUB

Date	Time (UTC)	Frequency or Band	Mode	Power	Station worked	Report sent	Report rec'd	QTH and Name	QSL sent	QSL rec'd
									☐	☐
									☐	☐
									☐	☐
									☐	☐
									☐	☐
									☐	☐
									☐	☐
									☐	☐
									☐	☐
									☐	☐

AMERICAN RADIO CLUB
Notes

AMERICAN RADIO CLUB

Date	Time (UTC)	Frequency or Band	Mode	Power	Station worked	Report sent	Report rec'd	QTH and Name	QSL sent	QSL rec'd
									☐	☐
									☐	☐
									☐	☐
									☐	☐
									☐	☐
									☐	☐
									☐	☐
									☐	☐
									☐	☐
									☐	☐

AMERICAN RADIO CLUB
Notes

AMERICAN RADIO CLUB

Date	Time (UTC)	Frequency or Band	Mode	Power	Station worked	Report sent	Report rec'd	QTH and Name	QSL sent	QSL rec'd
									☐	☐
									☐	☐
									☐	☐
									☐	☐
									☐	☐
									☐	☐
									☐	☐
									☐	☐
									☐	☐
									☐	☐

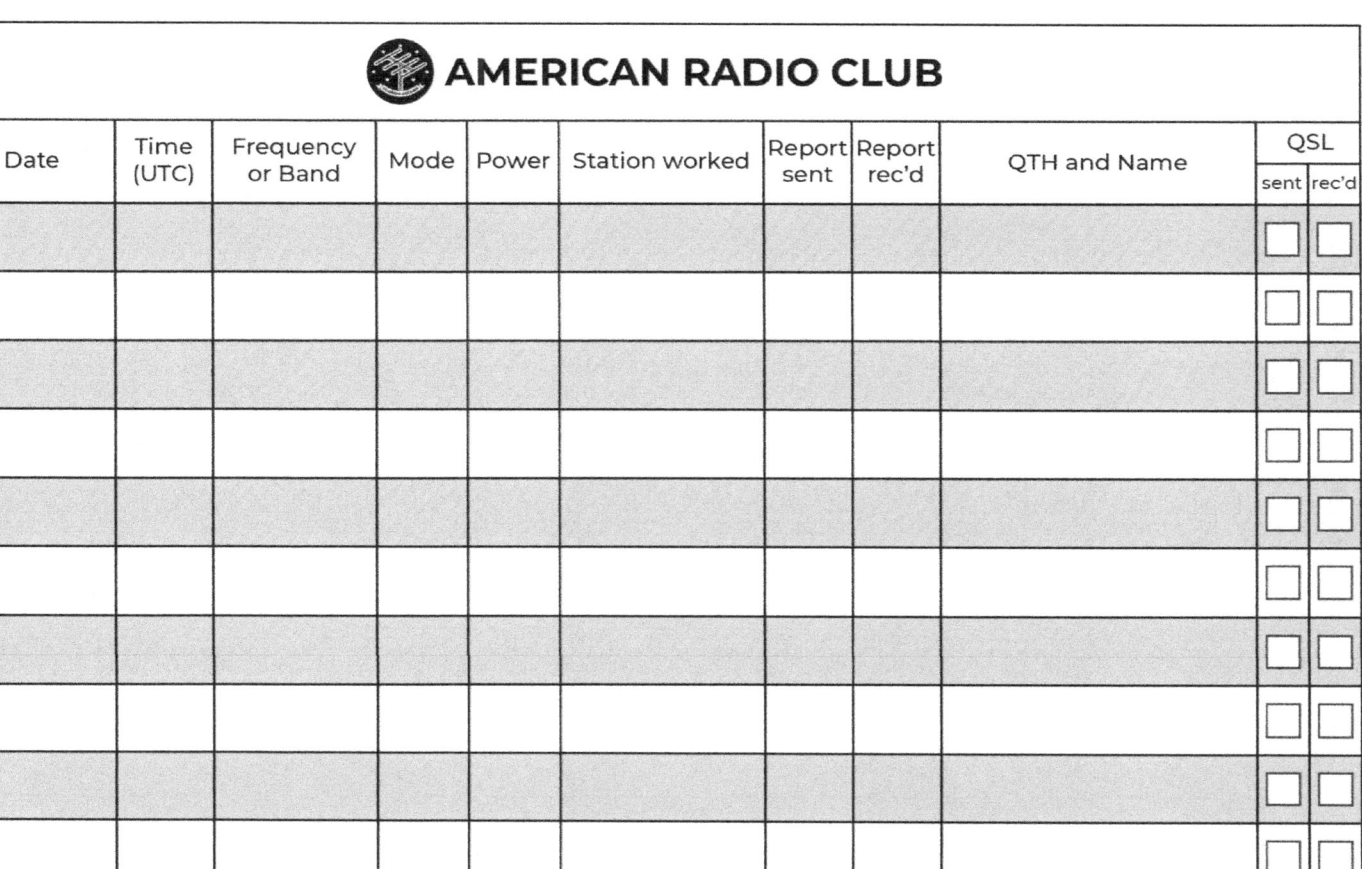

AMERICAN RADIO CLUB
Notes

AMERICAN RADIO CLUB

Date	Time (UTC)	Frequency or Band	Mode	Power	Station worked	Report sent	Report rec'd	QTH and Name	QSL sent	QSL rec'd
									☐	☐
									☐	☐
									☐	☐
									☐	☐
									☐	☐
									☐	☐
									☐	☐
									☐	☐
									☐	☐
									☐	☐

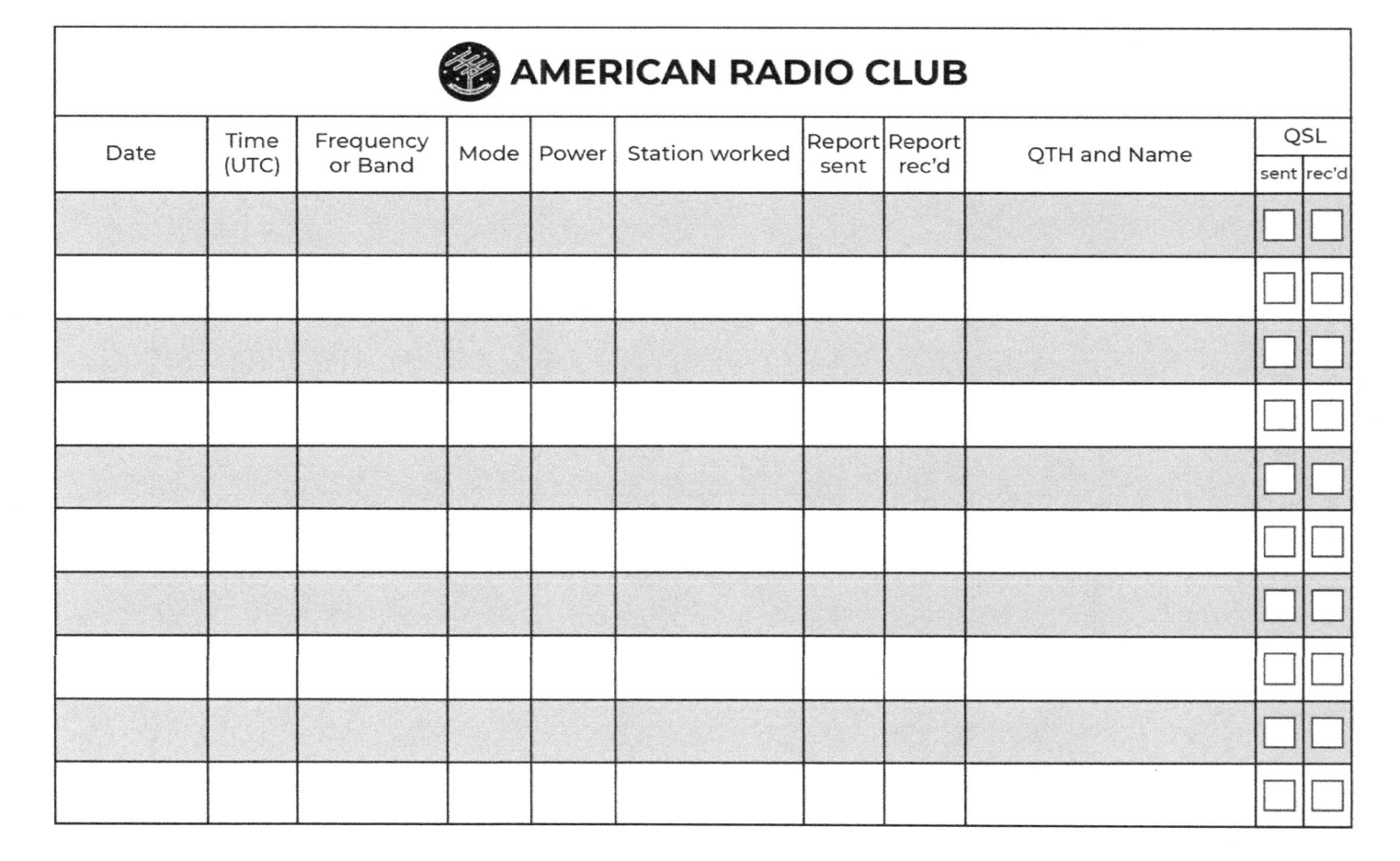

AMERICAN RADIO CLUB
Notes

AMERICAN RADIO CLUB

Date	Time (UTC)	Frequency or Band	Mode	Power	Station worked	Report sent	Report rec'd	QTH and Name	QSL sent	QSL rec'd
									☐	☐
									☐	☐
									☐	☐
									☐	☐
									☐	☐
									☐	☐
									☐	☐
									☐	☐
									☐	☐
									☐	☐

AMERICAN RADIO CLUB
Notes

AMERICAN RADIO CLUB

Date	Time (UTC)	Frequency or Band	Mode	Power	Station worked	Report sent	Report rec'd	QTH and Name	QSL sent	QSL rec'd
									☐	☐
									☐	☐
									☐	☐
									☐	☐
									☐	☐
									☐	☐
									☐	☐
									☐	☐
									☐	☐
									☐	☐

AMERICAN RADIO CLUB
Notes

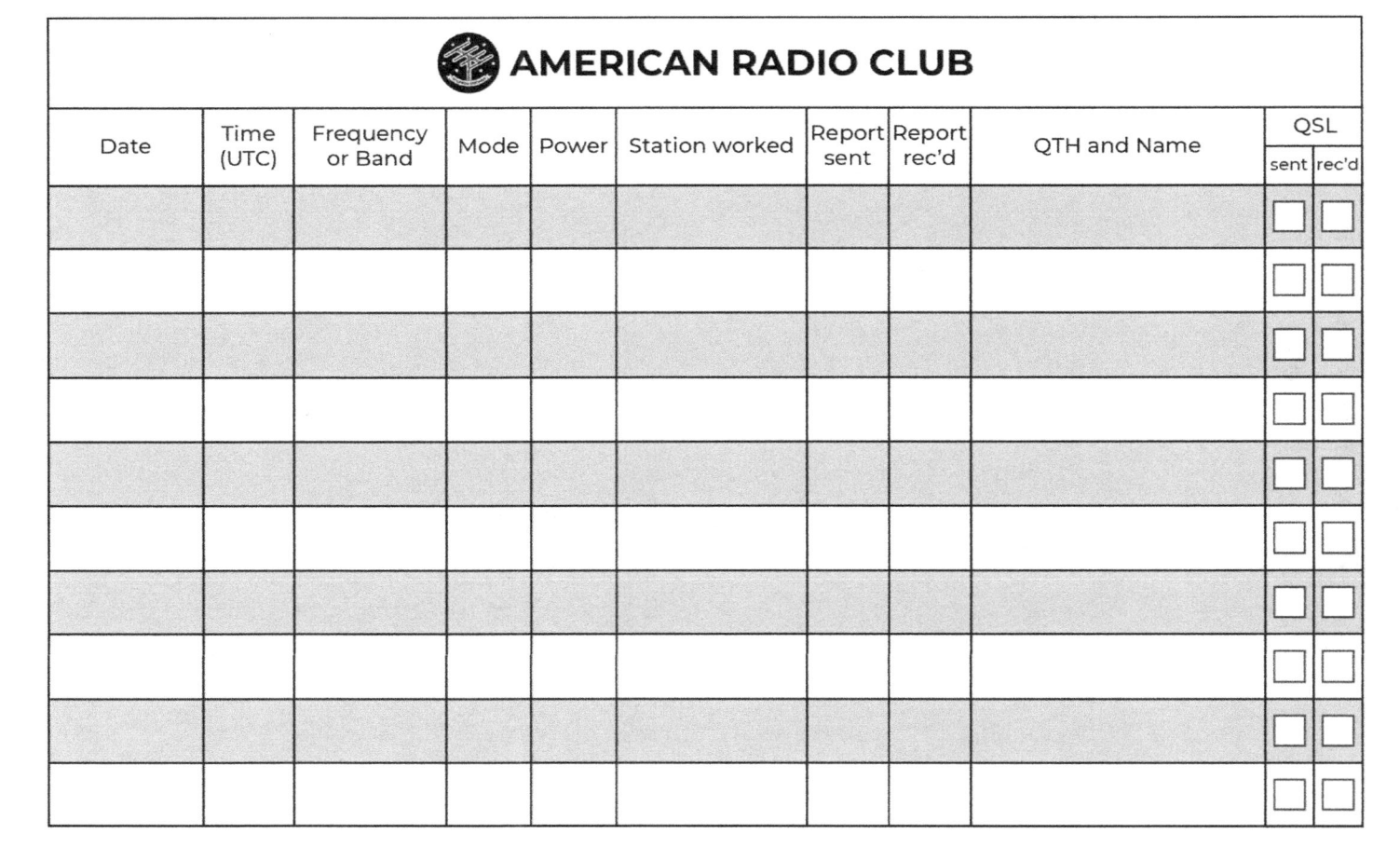

AMERICAN RADIO CLUB

Date	Time (UTC)	Frequency or Band	Mode	Power	Station worked	Report sent	Report rec'd	QTH and Name	QSL sent	QSL rec'd
									☐	☐
									☐	☐
									☐	☐
									☐	☐
									☐	☐
									☐	☐
									☐	☐
									☐	☐
									☐	☐
									☐	☐

AMERICAN RADIO CLUB
Notes

AMERICAN RADIO CLUB

Date	Time (UTC)	Frequency or Band	Mode	Power	Station worked	Report sent	Report rec'd	QTH and Name	QSL sent	QSL rec'd
									☐	☐
									☐	☐
									☐	☐
									☐	☐
									☐	☐
									☐	☐
									☐	☐
									☐	☐
									☐	☐
									☐	☐

AMERICAN RADIO CLUB
Notes

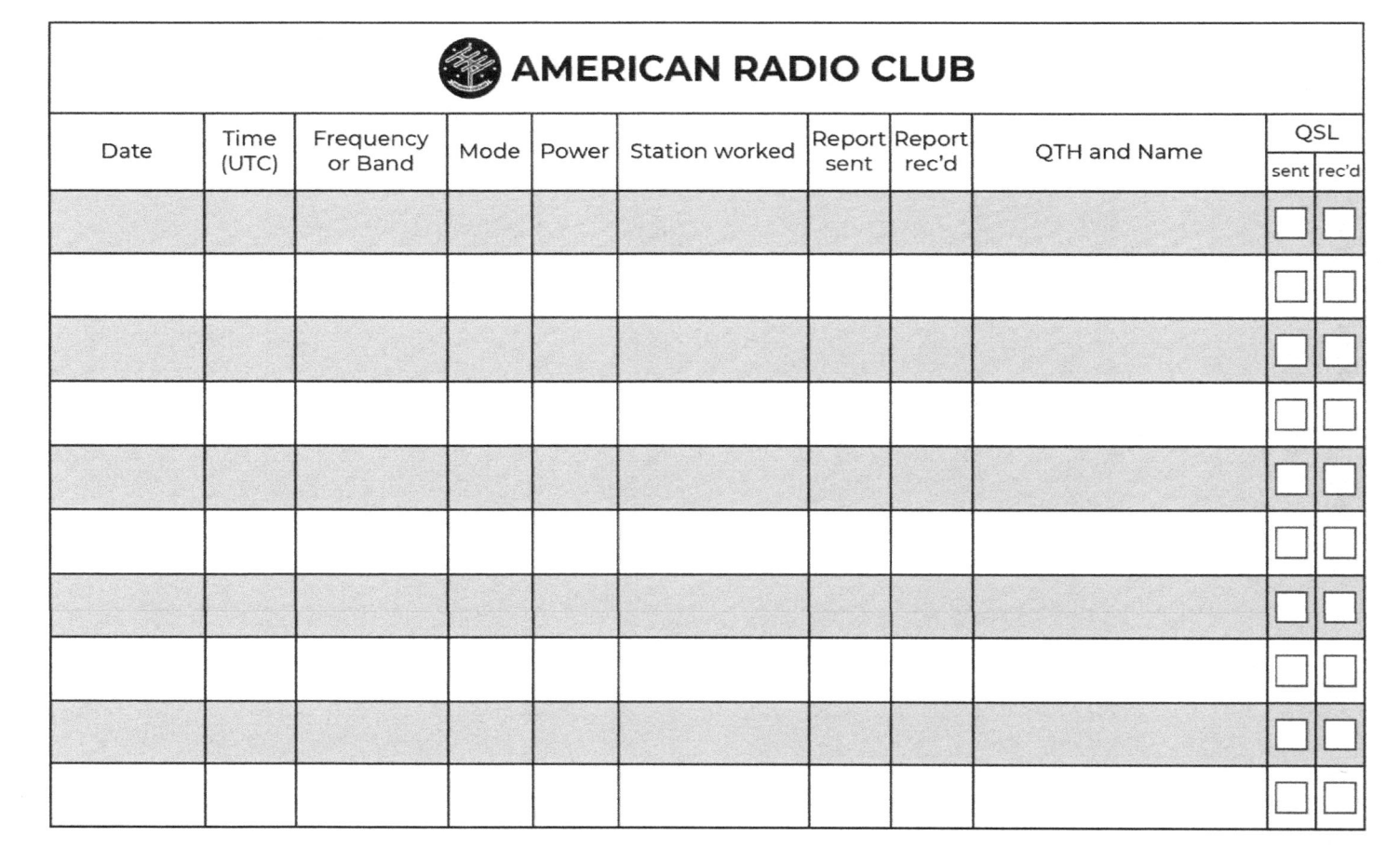

AMERICAN RADIO CLUB

Date	Time (UTC)	Frequency or Band	Mode	Power	Station worked	Report sent	Report rec'd	QTH and Name	QSL sent	QSL rec'd
									☐	☐
									☐	☐
									☐	☐
									☐	☐
									☐	☐
									☐	☐
									☐	☐
									☐	☐
									☐	☐
									☐	☐

AMERICAN RADIO CLUB
Notes

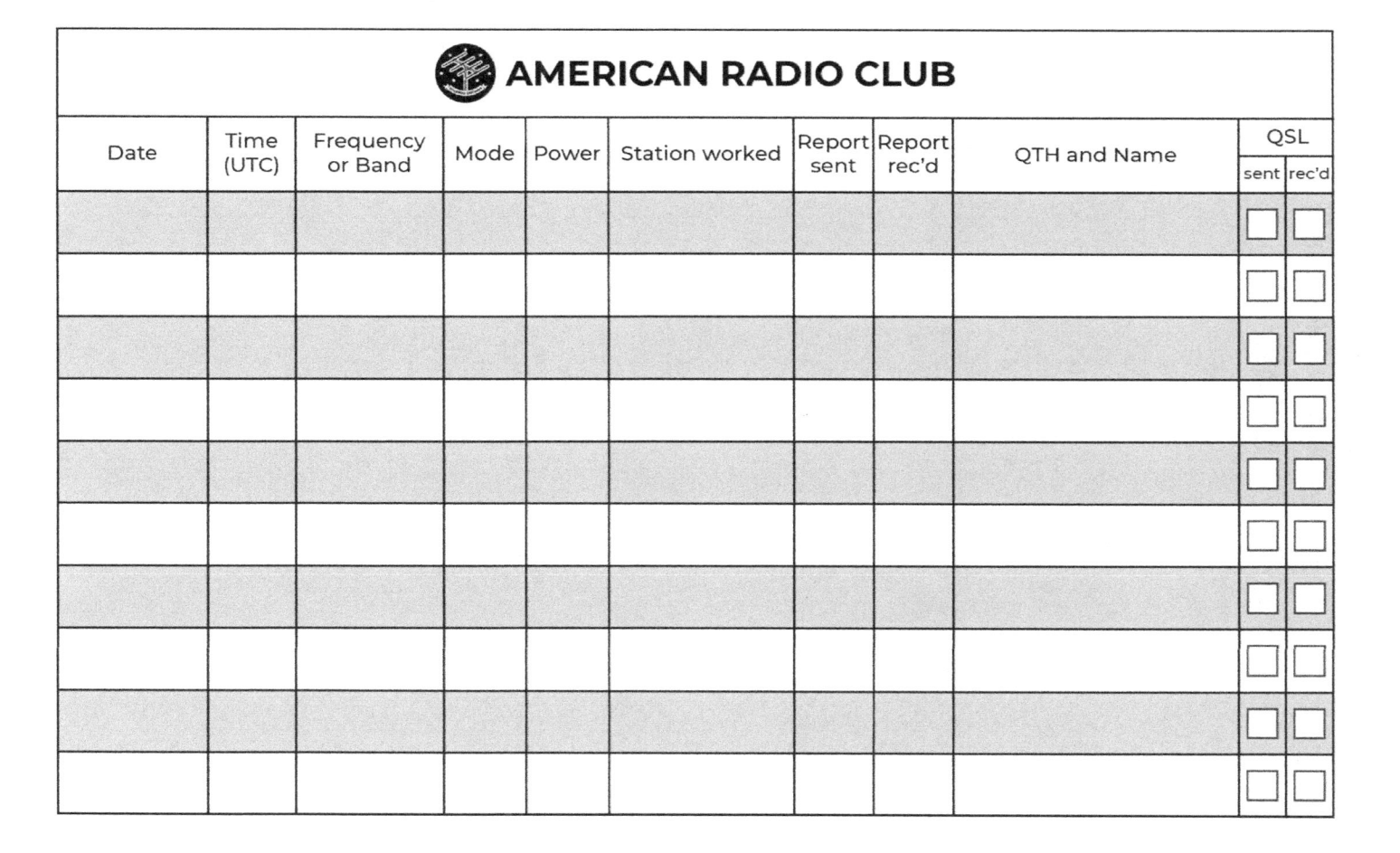

AMERICAN RADIO CLUB

Date	Time (UTC)	Frequency or Band	Mode	Power	Station worked	Report sent	Report rec'd	QTH and Name	QSL sent	QSL rec'd
									☐	☐
									☐	☐
									☐	☐
									☐	☐
									☐	☐
									☐	☐
									☐	☐
									☐	☐
									☐	☐
									☐	☐

AMERICAN RADIO CLUB
Notes

AMERICAN RADIO CLUB

Date	Time (UTC)	Frequency or Band	Mode	Power	Station worked	Report sent	Report rec'd	QTH and Name	QSL sent	QSL rec'd
									☐	☐
									☐	☐
									☐	☐
									☐	☐
									☐	☐
									☐	☐
									☐	☐
									☐	☐
									☐	☐
									☐	☐

AMERICAN RADIO CLUB
Notes

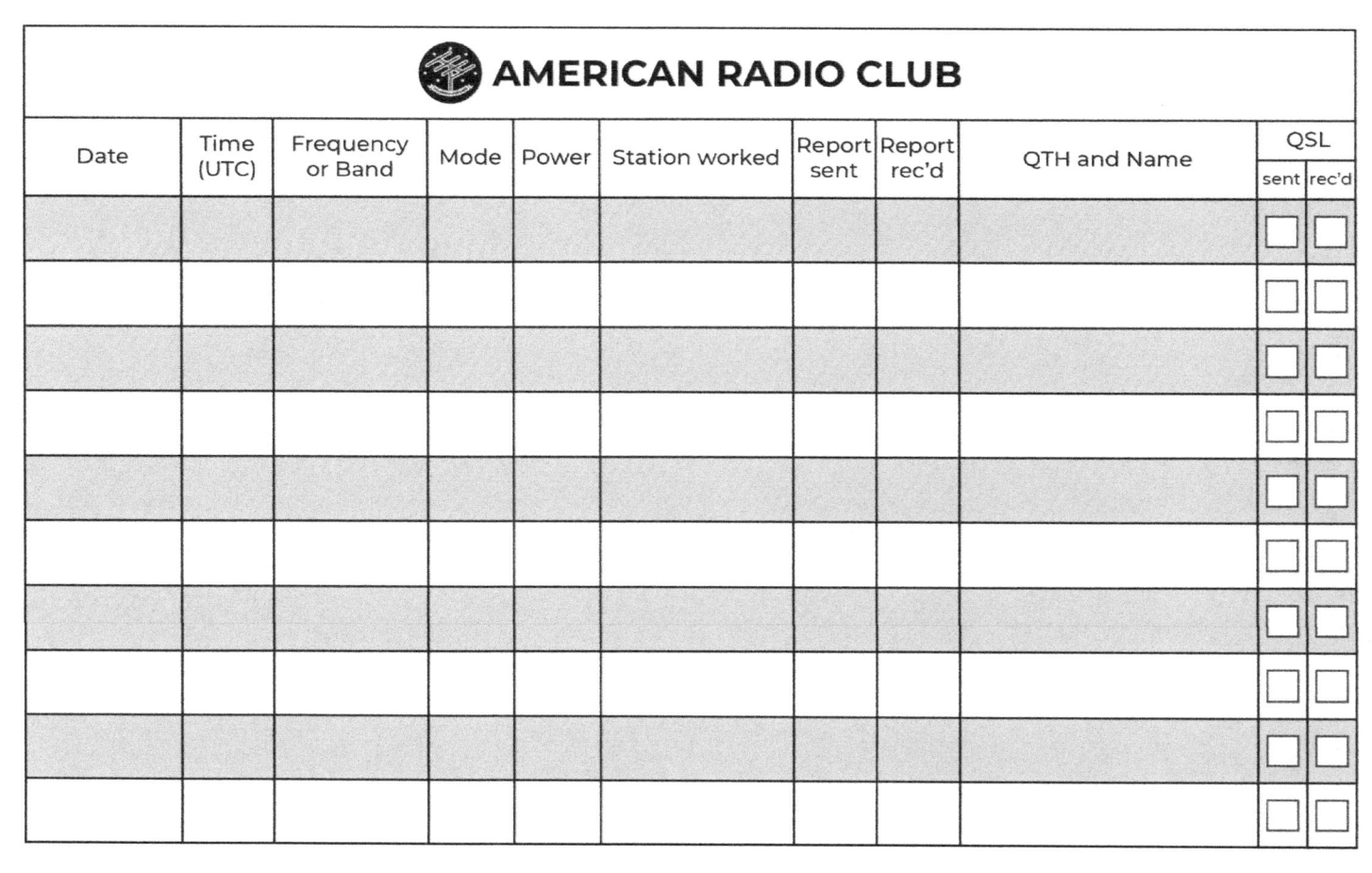

AMERICAN RADIO CLUB

Date	Time (UTC)	Frequency or Band	Mode	Power	Station worked	Report sent	Report rec'd	QTH and Name	QSL sent	QSL rec'd
									☐	☐
									☐	☐
									☐	☐
									☐	☐
									☐	☐
									☐	☐
									☐	☐
									☐	☐
									☐	☐
									☐	☐

AMERICAN RADIO CLUB
Notes

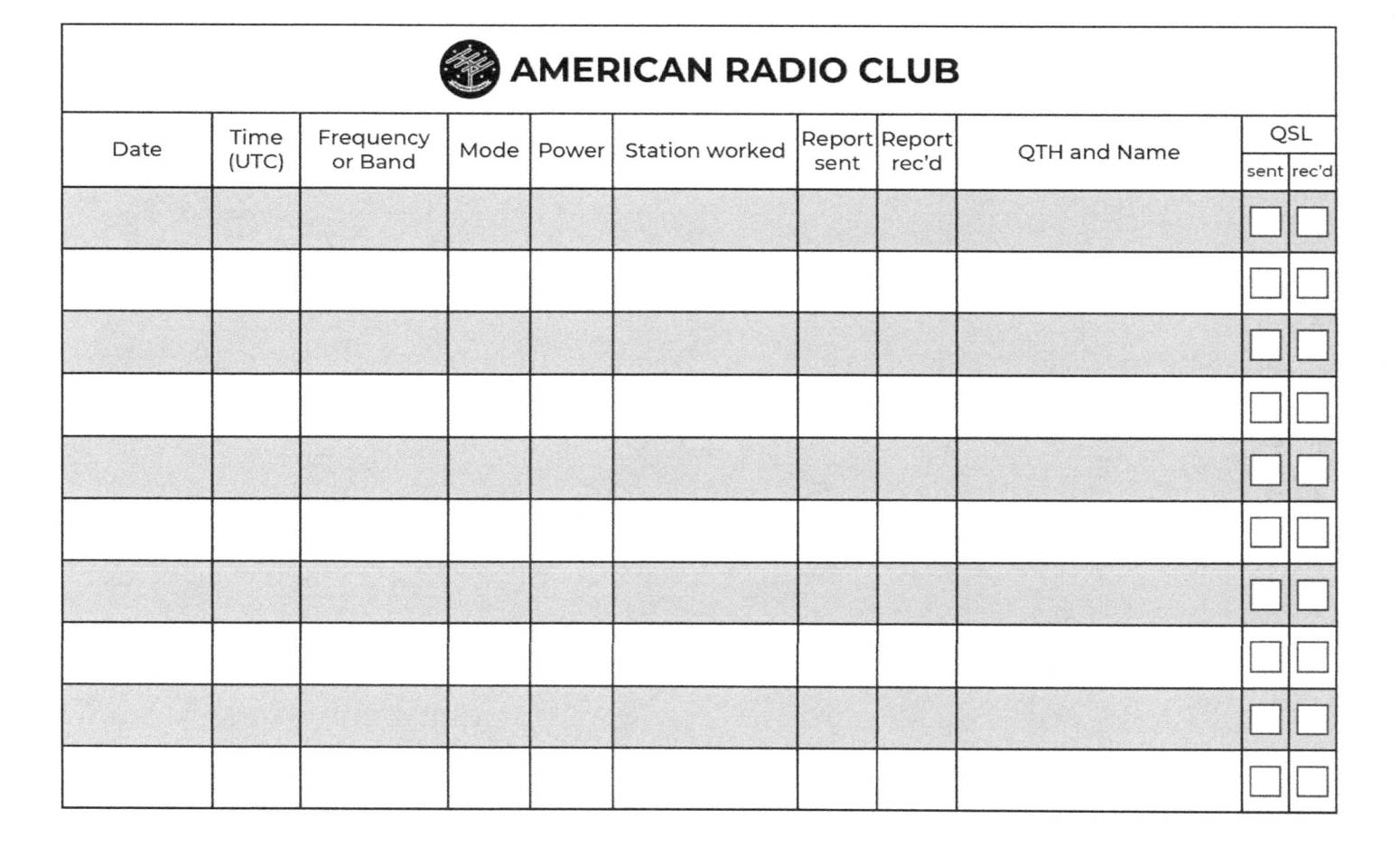

AMERICAN RADIO CLUB

Date	Time (UTC)	Frequency or Band	Mode	Power	Station worked	Report sent	Report rec'd	QTH and Name	QSL sent	QSL rec'd
									☐	☐
									☐	☐
									☐	☐
									☐	☐
									☐	☐
									☐	☐
									☐	☐
									☐	☐
									☐	☐
									☐	☐

AMERICAN RADIO CLUB
Notes

AMERICAN RADIO CLUB

Date	Time (UTC)	Frequency or Band	Mode	Power	Station worked	Report sent	Report rec'd	QTH and Name	QSL sent	QSL rec'd
									☐	☐
									☐	☐
									☐	☐
									☐	☐
									☐	☐
									☐	☐
									☐	☐
									☐	☐
									☐	☐
									☐	☐

AMERICAN RADIO CLUB
Notes

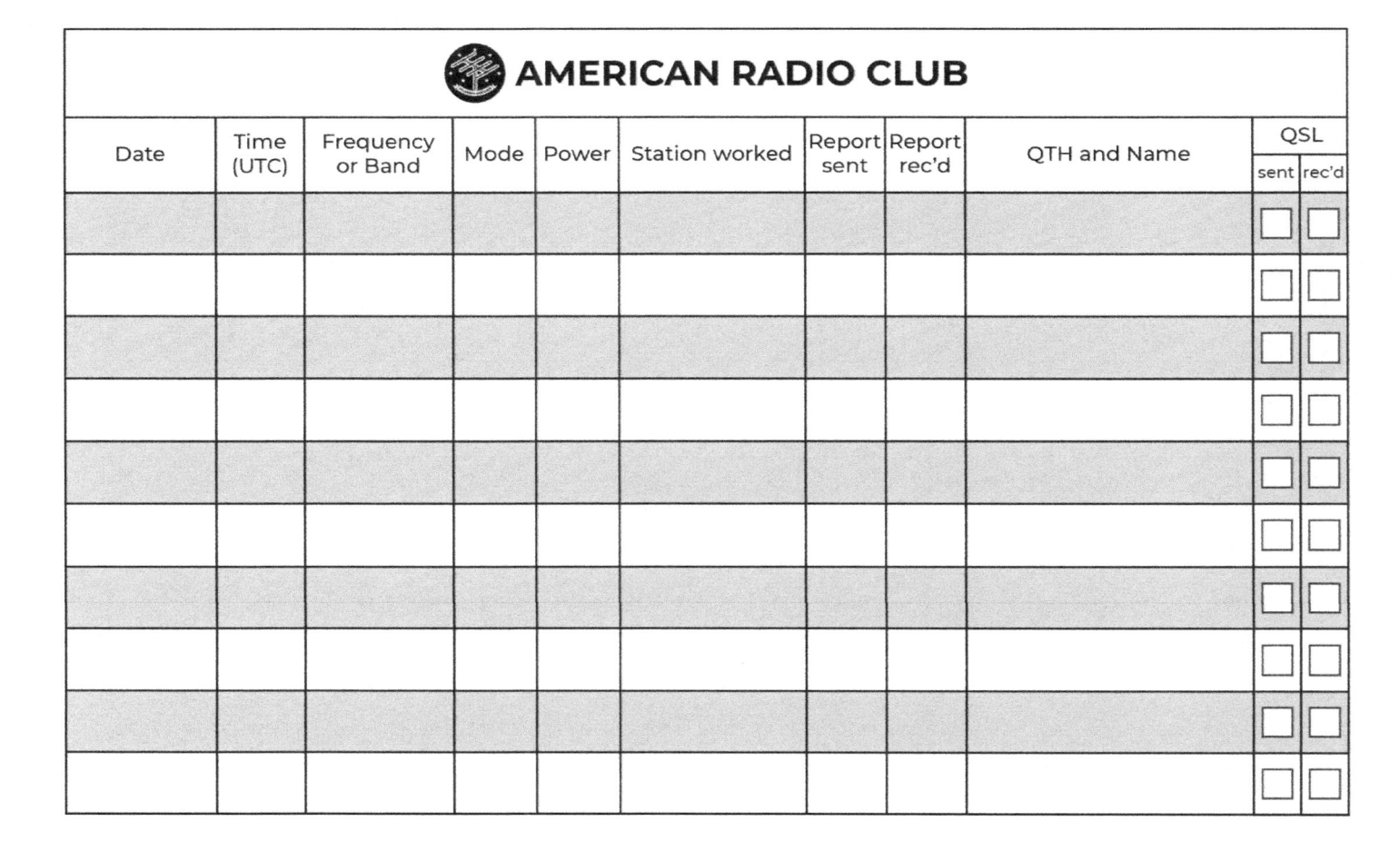

AMERICAN RADIO CLUB

Date	Time (UTC)	Frequency or Band	Mode	Power	Station worked	Report sent	Report rec'd	QTH and Name	QSL sent	QSL rec'd
									☐	☐
									☐	☐
									☐	☐
									☐	☐
									☐	☐
									☐	☐
									☐	☐
									☐	☐
									☐	☐
									☐	☐

AMERICAN RADIO CLUB
Notes

AMERICAN RADIO CLUB

Date	Time (UTC)	Frequency or Band	Mode	Power	Station worked	Report sent	Report rec'd	QTH and Name	QSL sent	QSL rec'd
									☐	☐
									☐	☐
									☐	☐
									☐	☐
									☐	☐
									☐	☐
									☐	☐
									☐	☐
									☐	☐
									☐	☐

AMERICAN RADIO CLUB
Notes

AMERICAN RADIO CLUB

Date	Time (UTC)	Frequency or Band	Mode	Power	Station worked	Report sent	Report rec'd	QTH and Name	QSL sent	QSL rec'd
									☐	☐
									☐	☐
									☐	☐
									☐	☐
									☐	☐
									☐	☐
									☐	☐
									☐	☐
									☐	☐
									☐	☐

AMERICAN RADIO CLUB
Notes

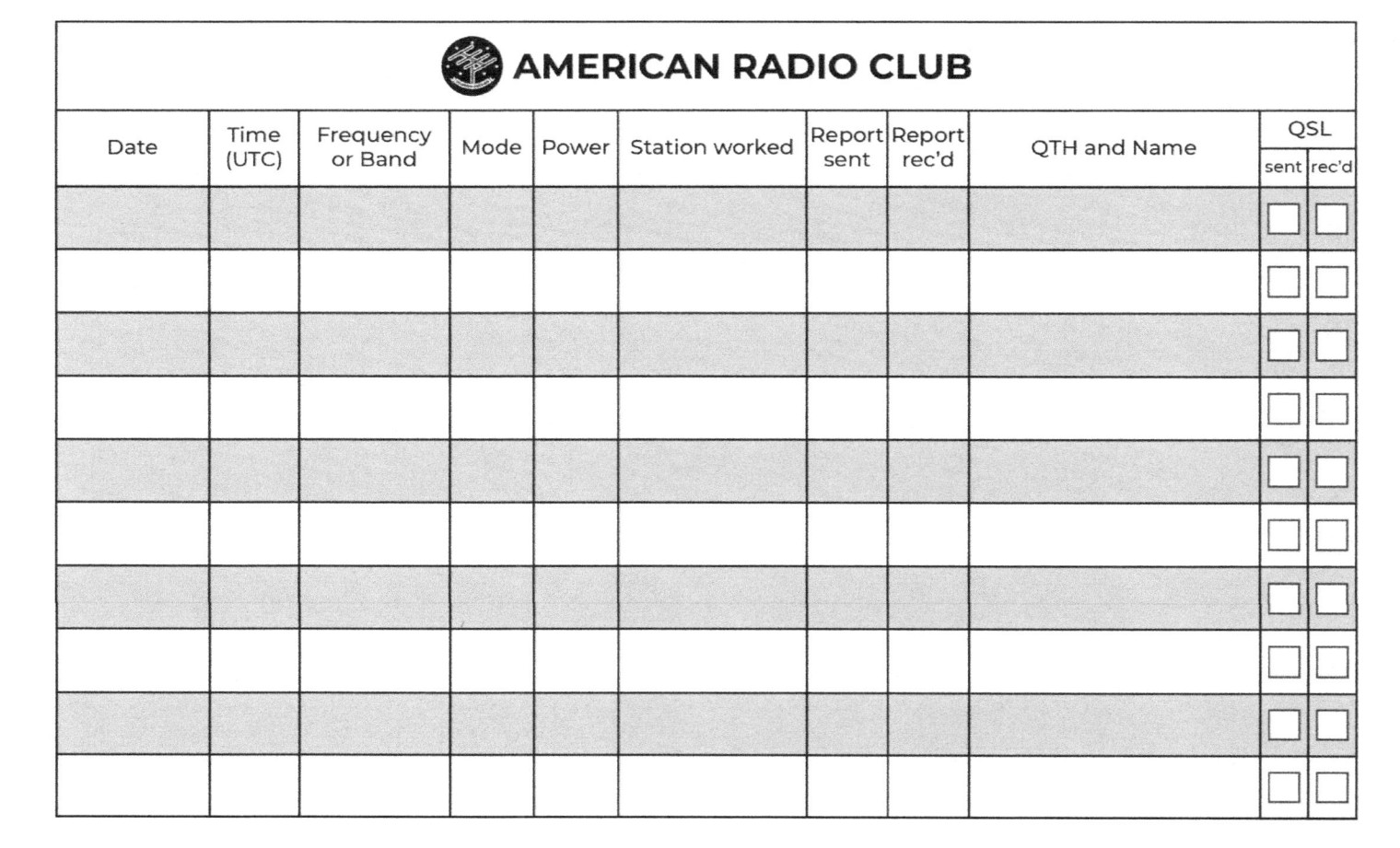

AMERICAN RADIO CLUB

Date	Time (UTC)	Frequency or Band	Mode	Power	Station worked	Report sent	Report rec'd	QTH and Name	QSL sent	QSL rec'd
									☐	☐
									☐	☐
									☐	☐
									☐	☐
									☐	☐
									☐	☐
									☐	☐
									☐	☐
									☐	☐
									☐	☐

AMERICAN RADIO CLUB
Notes

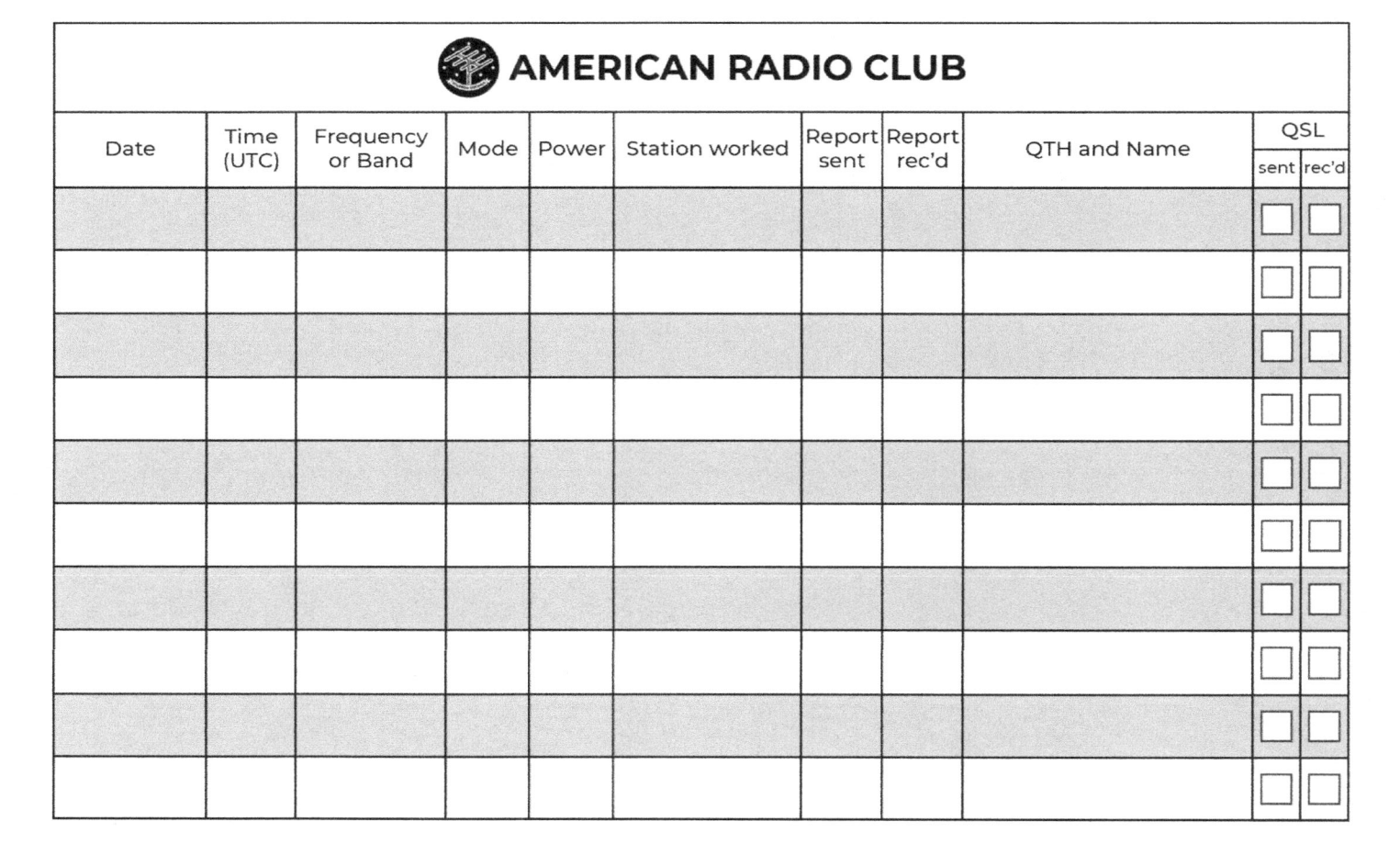

AMERICAN RADIO CLUB

Date	Time (UTC)	Frequency or Band	Mode	Power	Station worked	Report sent	Report rec'd	QTH and Name	QSL sent	QSL rec'd
									☐	☐
									☐	☐
									☐	☐
									☐	☐
									☐	☐
									☐	☐
									☐	☐
									☐	☐
									☐	☐
									☐	☐

AMERICAN RADIO CLUB
Notes

AMERICAN RADIO CLUB

Date	Time (UTC)	Frequency or Band	Mode	Power	Station worked	Report sent	Report rec'd	QTH and Name	QSL sent	QSL rec'd
									☐	☐
									☐	☐
									☐	☐
									☐	☐
									☐	☐
									☐	☐
									☐	☐
									☐	☐
									☐	☐
									☐	☐

AMERICAN RADIO CLUB
Notes

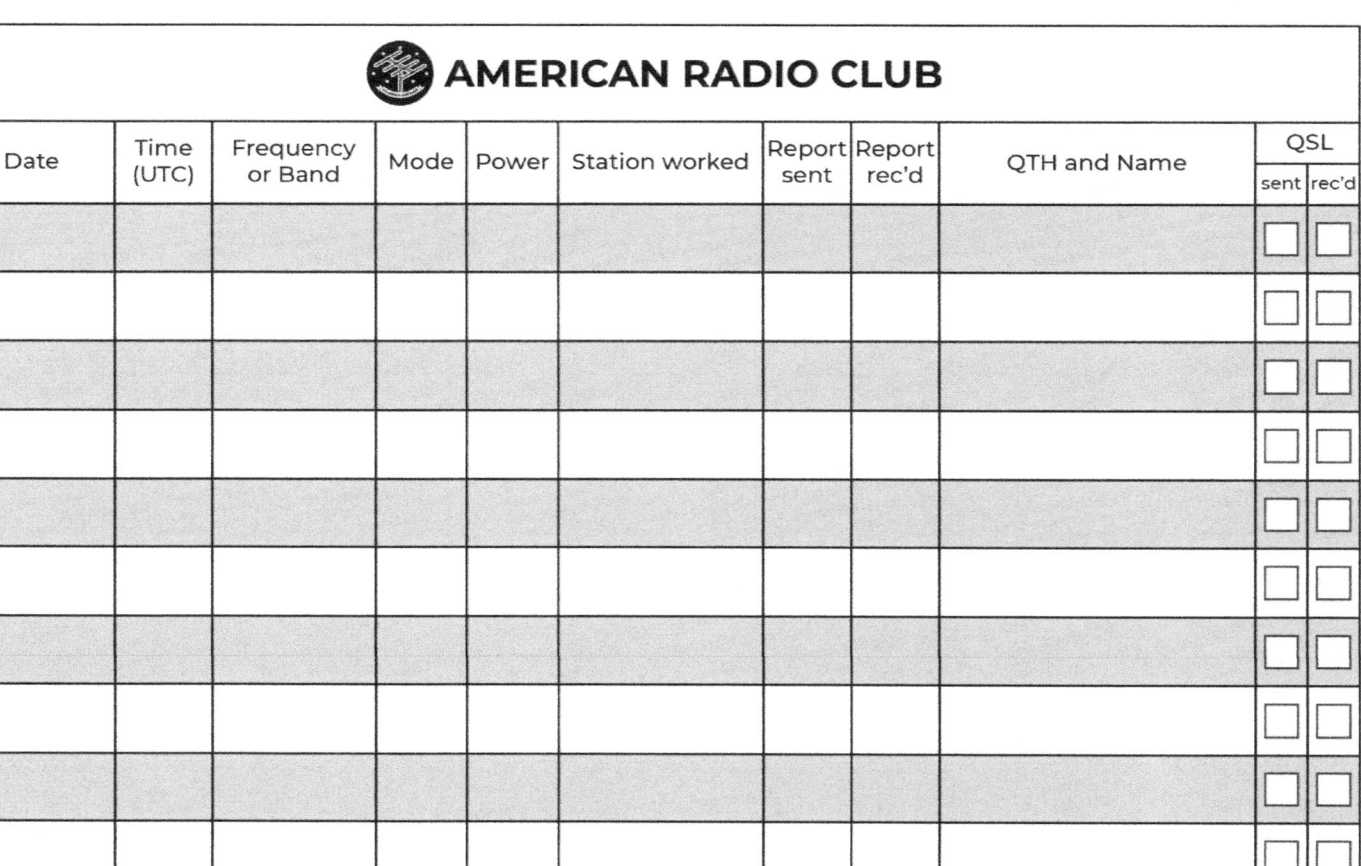

AMERICAN RADIO CLUB

Date	Time (UTC)	Frequency or Band	Mode	Power	Station worked	Report sent	Report rec'd	QTH and Name	QSL sent	QSL rec'd
									☐	☐
									☐	☐
									☐	☐
									☐	☐
									☐	☐
									☐	☐
									☐	☐
									☐	☐
									☐	☐
									☐	☐

AMERICAN RADIO CLUB
Notes

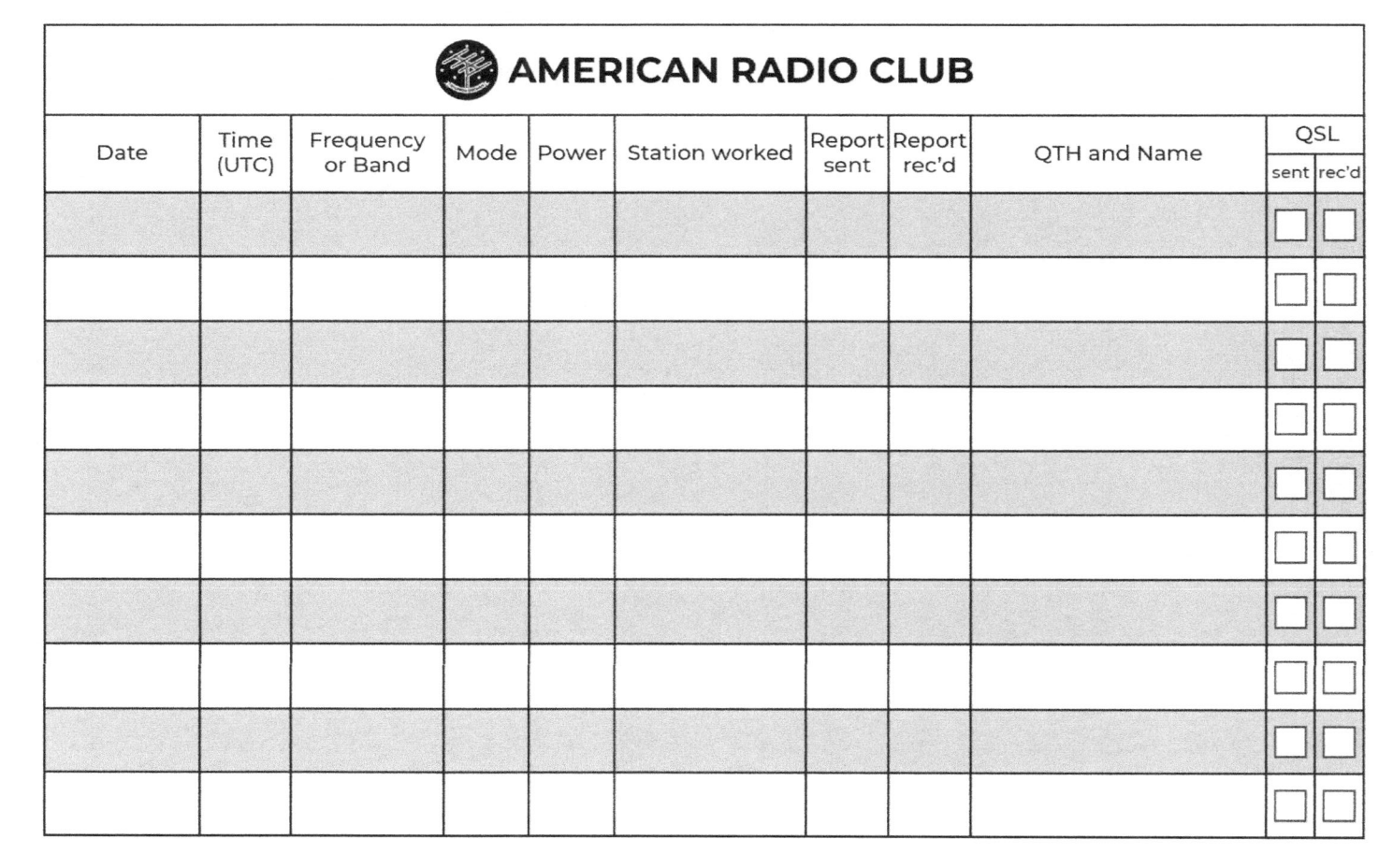

AMERICAN RADIO CLUB

Date	Time (UTC)	Frequency or Band	Mode	Power	Station worked	Report sent	Report rec'd	QTH and Name	QSL sent	QSL rec'd
									☐	☐
									☐	☐
									☐	☐
									☐	☐
									☐	☐
									☐	☐
									☐	☐
									☐	☐
									☐	☐
									☐	☐

AMERICAN RADIO CLUB
Notes

AMERICAN RADIO CLUB

Date	Time (UTC)	Frequency or Band	Mode	Power	Station worked	Report sent	Report rec'd	QTH and Name	QSL sent	QSL rec'd
									☐	☐
									☐	☐
									☐	☐
									☐	☐
									☐	☐
									☐	☐
									☐	☐
									☐	☐
									☐	☐
									☐	☐

AMERICAN RADIO CLUB
Notes

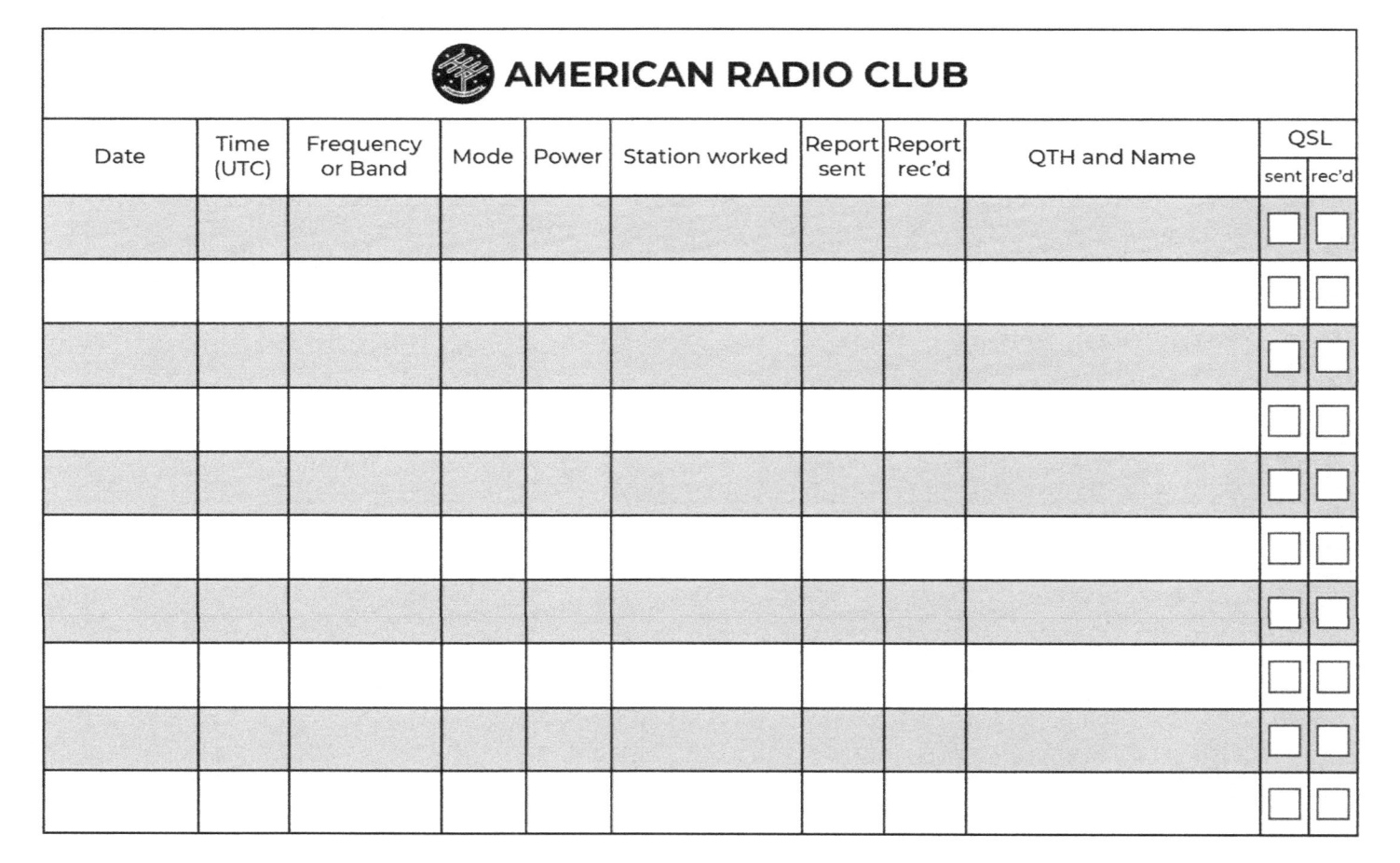

AMERICAN RADIO CLUB

Date	Time (UTC)	Frequency or Band	Mode	Power	Station worked	Report sent	Report rec'd	QTH and Name	QSL sent	QSL rec'd
									☐	☐
									☐	☐
									☐	☐
									☐	☐
									☐	☐
									☐	☐
									☐	☐
									☐	☐
									☐	☐
									☐	☐

AMERICAN RADIO CLUB
Notes

AMERICAN RADIO CLUB

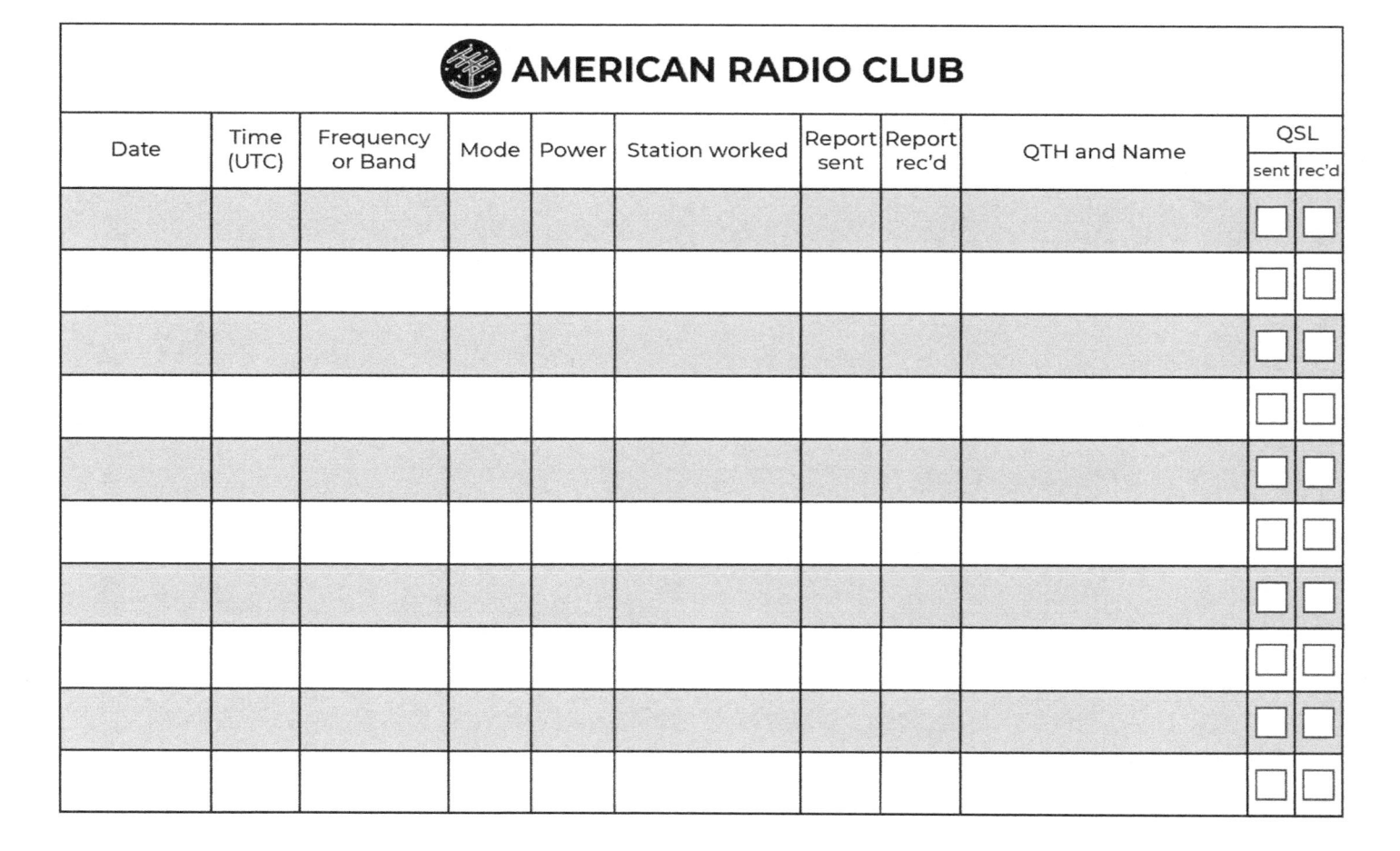

Date	Time (UTC)	Frequency or Band	Mode	Power	Station worked	Report sent	Report rec'd	QTH and Name	QSL sent	QSL rec'd
									☐	☐
									☐	☐
									☐	☐
									☐	☐
									☐	☐
									☐	☐
									☐	☐
									☐	☐
									☐	☐
									☐	☐

AMERICAN RADIO CLUB
Notes

AMERICAN RADIO CLUB

Date	Time (UTC)	Frequency or Band	Mode	Power	Station worked	Report sent	Report rec'd	QTH and Name	QSL sent	QSL rec'd
									☐	☐
									☐	☐
									☐	☐
									☐	☐
									☐	☐
									☐	☐
									☐	☐
									☐	☐
									☐	☐
									☐	☐

AMERICAN RADIO CLUB
Notes

AMERICAN RADIO CLUB

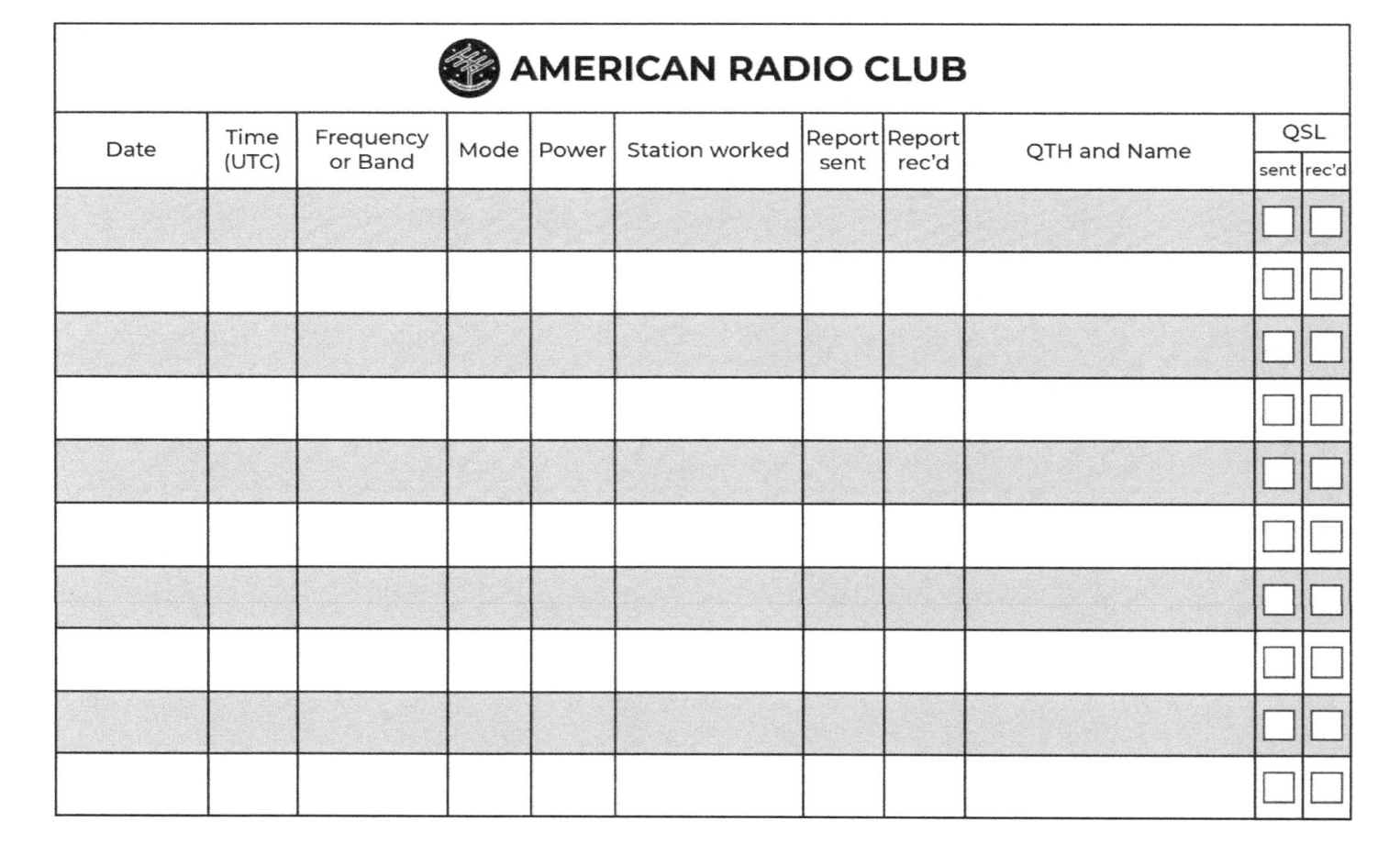

Date	Time (UTC)	Frequency or Band	Mode	Power	Station worked	Report sent	Report rec'd	QTH and Name	QSL sent	QSL rec'd
									☐	☐
									☐	☐
									☐	☐
									☐	☐
									☐	☐
									☐	☐
									☐	☐
									☐	☐
									☐	☐
									☐	☐

AMERICAN RADIO CLUB
Notes

AMERICAN RADIO CLUB

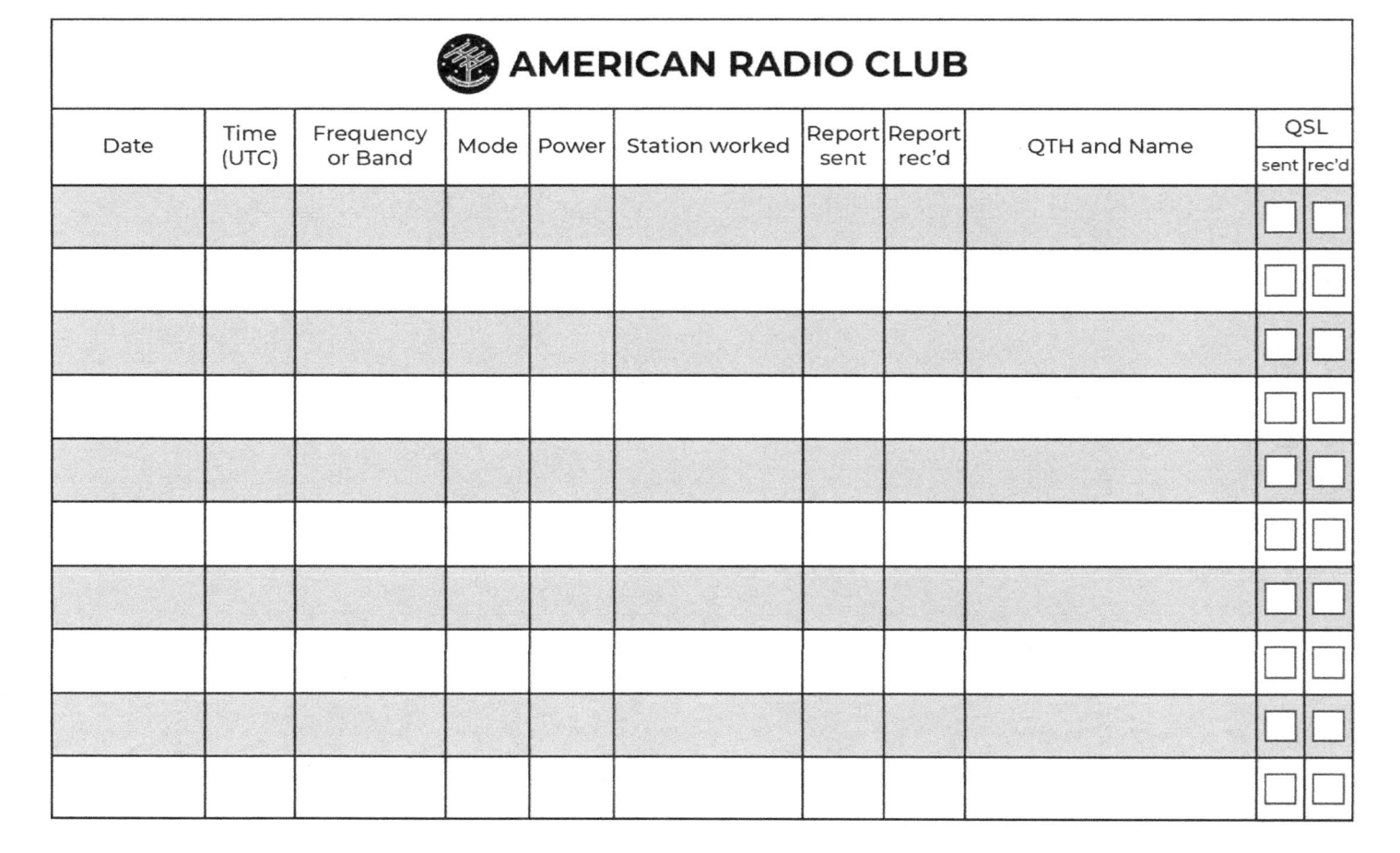

Date	Time (UTC)	Frequency or Band	Mode	Power	Station worked	Report sent	Report rec'd	QTH and Name	QSL sent	QSL rec'd
									☐	☐
									☐	☐
									☐	☐
									☐	☐
									☐	☐
									☐	☐
									☐	☐
									☐	☐
									☐	☐
									☐	☐

AMERICAN RADIO CLUB
Notes

AMERICAN RADIO CLUB

Date	Time (UTC)	Frequency or Band	Mode	Power	Station worked	Report sent	Report rec'd	QTH and Name	QSL sent	QSL rec'd
									☐	☐
									☐	☐
									☐	☐
									☐	☐
									☐	☐
									☐	☐
									☐	☐
									☐	☐
									☐	☐
									☐	☐

AMERICAN RADIO CLUB
Notes

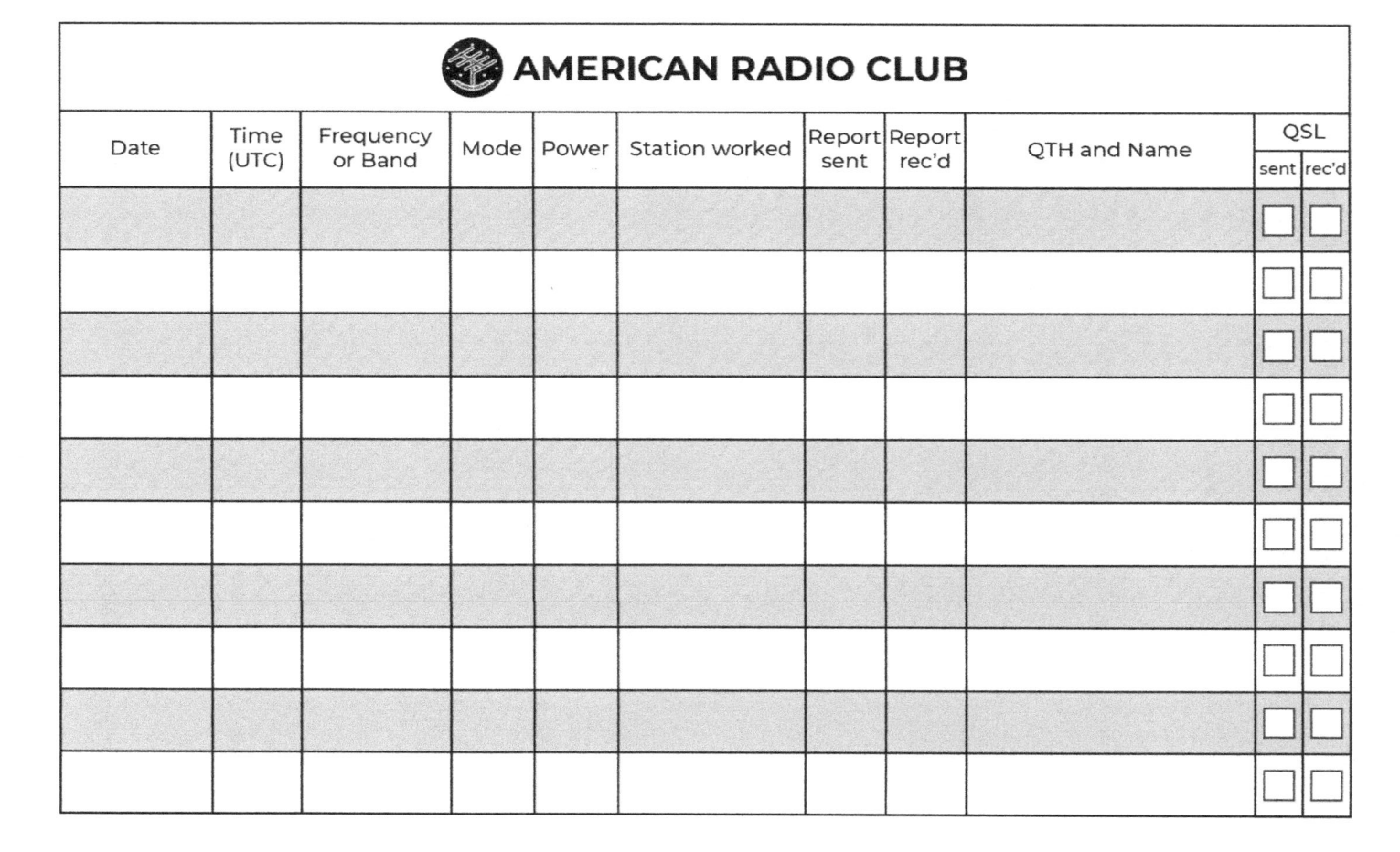

AMERICAN RADIO CLUB

Date	Time (UTC)	Frequency or Band	Mode	Power	Station worked	Report sent	Report rec'd	QTH and Name	QSL sent	QSL rec'd
									☐	☐
									☐	☐
									☐	☐
									☐	☐
									☐	☐
									☐	☐
									☐	☐
									☐	☐
									☐	☐
									☐	☐

AMERICAN RADIO CLUB
Notes

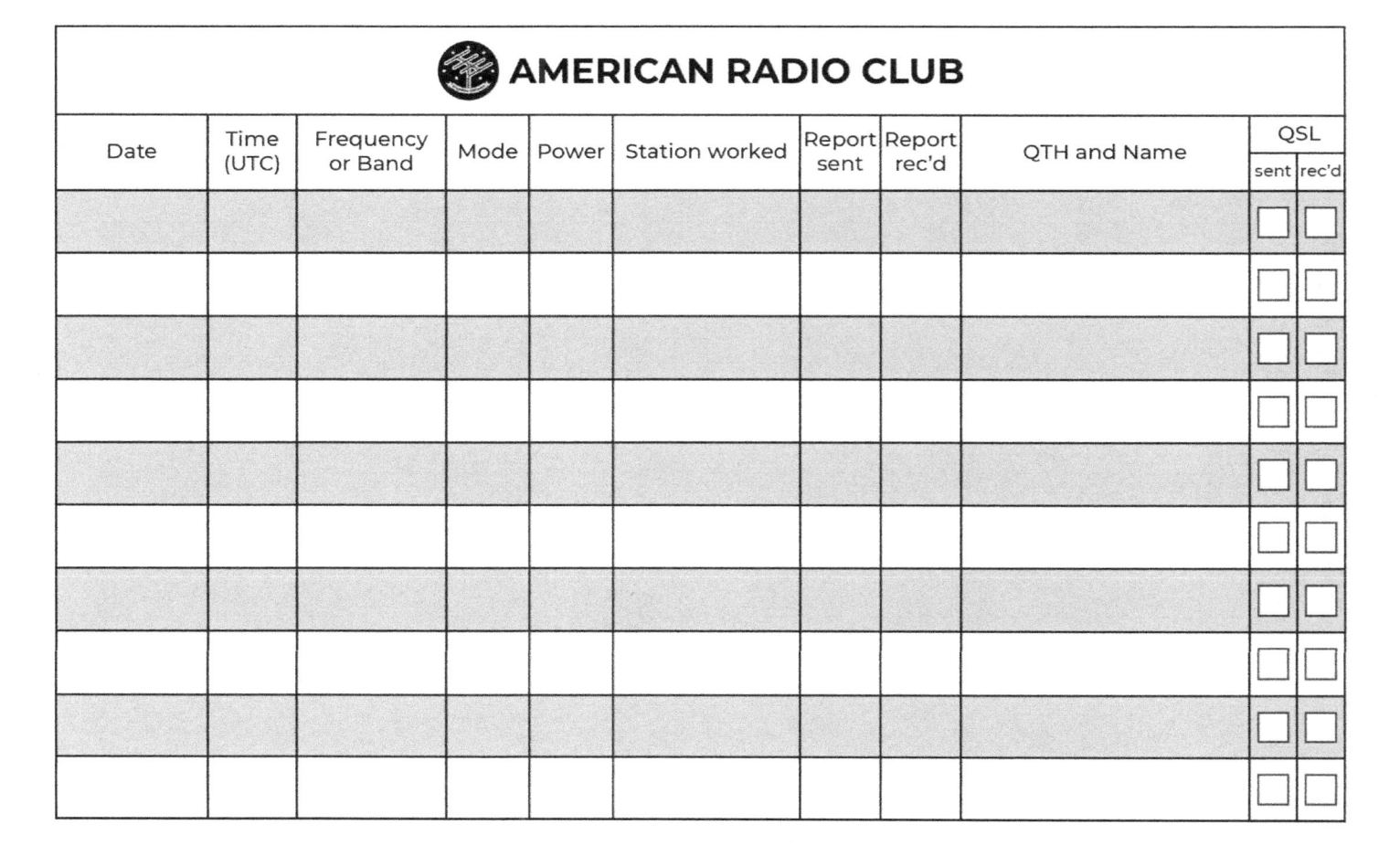

AMERICAN RADIO CLUB

Date	Time (UTC)	Frequency or Band	Mode	Power	Station worked	Report sent	Report rec'd	QTH and Name	QSL sent	QSL rec'd
									☐	☐
									☐	☐
									☐	☐
									☐	☐
									☐	☐
									☐	☐
									☐	☐
									☐	☐
									☐	☐
									☐	☐

AMERICAN RADIO CLUB
Notes

AMERICAN RADIO CLUB

Date	Time (UTC)	Frequency or Band	Mode	Power	Station worked	Report sent	Report rec'd	QTH and Name	QSL sent	QSL rec'd
									☐	☐
									☐	☐
									☐	☐
									☐	☐
									☐	☐
									☐	☐
									☐	☐
									☐	☐
									☐	☐
									☐	☐

AMERICAN RADIO CLUB
Notes

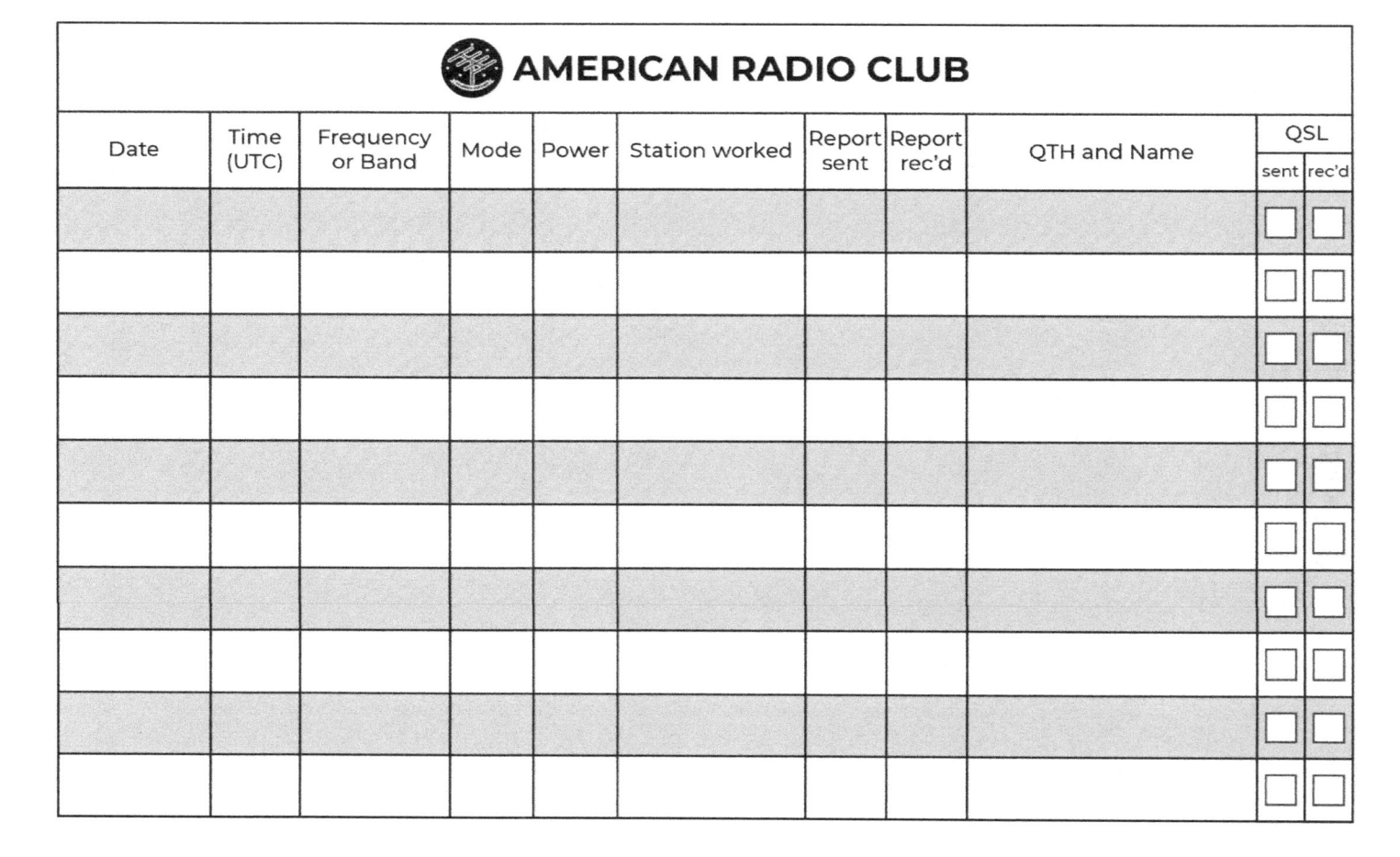

AMERICAN RADIO CLUB

Date	Time (UTC)	Frequency or Band	Mode	Power	Station worked	Report sent	Report rec'd	QTH and Name	QSL sent	QSL rec'd
									☐	☐
									☐	☐
									☐	☐
									☐	☐
									☐	☐
									☐	☐
									☐	☐
									☐	☐
									☐	☐
									☐	☐

AMERICAN RADIO CLUB
Notes

AMERICAN RADIO CLUB

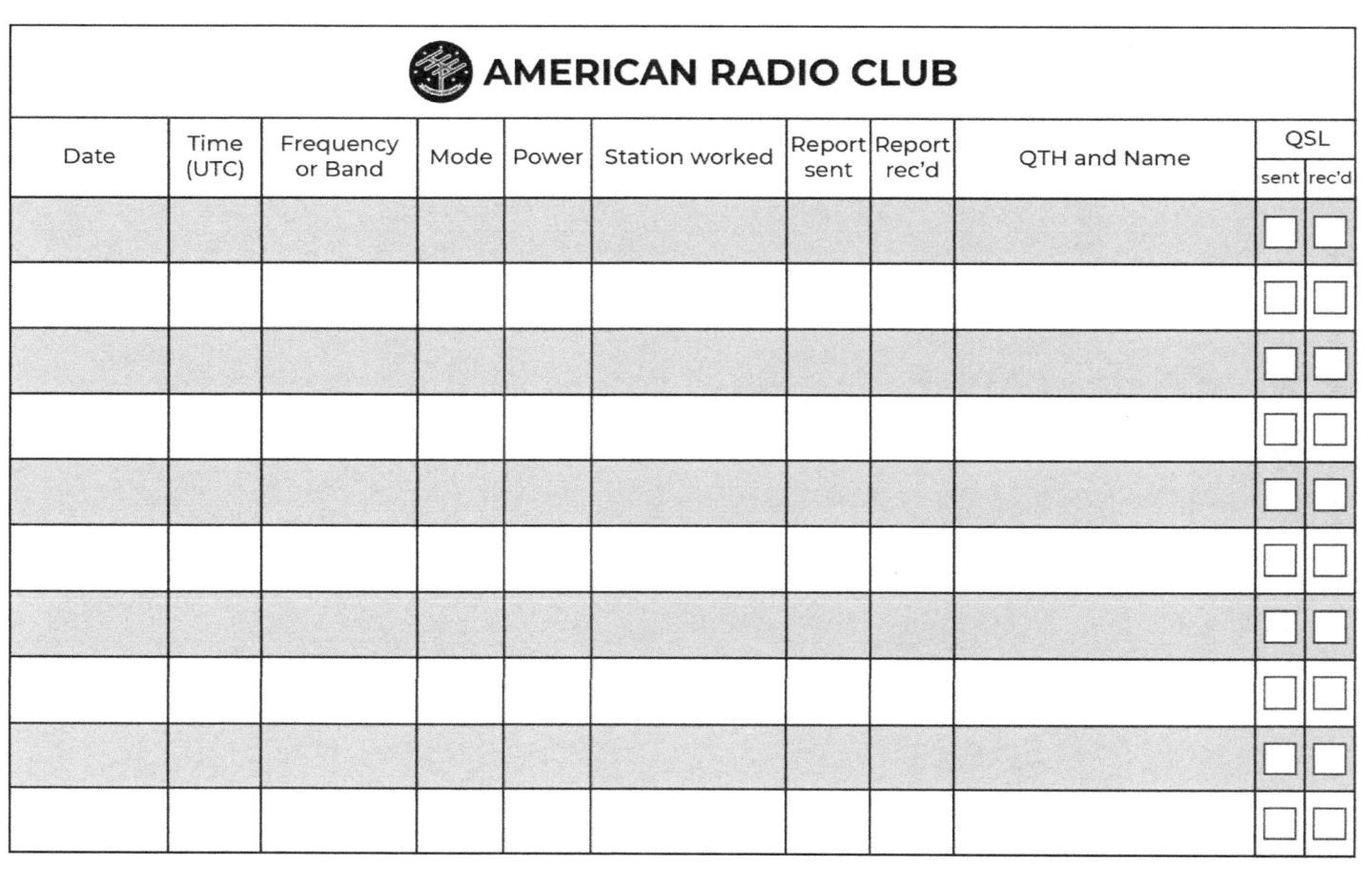

Date	Time (UTC)	Frequency or Band	Mode	Power	Station worked	Report sent	Report rec'd	QTH and Name	QSL sent	QSL rec'd
									☐	☐
									☐	☐
									☐	☐
									☐	☐
									☐	☐
									☐	☐
									☐	☐
									☐	☐
									☐	☐
									☐	☐

AMERICAN RADIO CLUB
Notes

… AMERICAN RADIO CLUB

Date	Time (UTC)	Frequency or Band	Mode	Power	Station worked	Report sent	Report rec'd	QTH and Name	QSL sent	QSL rec'd
									☐	☐
									☐	☐
									☐	☐
									☐	☐
									☐	☐
									☐	☐
									☐	☐
									☐	☐
									☐	☐
									☐	☐

AMERICAN RADIO CLUB
Notes

Resources

Our website: AmericanRadioClub.com

Facebook: facebook.com/AmerRadioClub

Reddit: reddit.com/user/AmericanRadioClub

Twitter: twitter.com/AmerRadioClub

YouTube: youtube.com/c/AmericanRadioClub

Instagram: instagram.com/american.radioclub

TikTok: tiktok.com/@americanradioclub

LinkedIn:
linkedin.com/company/AmericanRadioClub

Federal Communications Commission:
FCC.gov

Finding a repeater: repeaterbook.com

Phonetic alphabet:

Phonetic Alphabet

- **A**lpha
- **B**ravo
- **C**harlie
- **D**elta
- **E**cho
- **F**oxtrot
- **G**olf
- **H**otel
- **I**ndia
- **J**uliet
- **K**ilo
- **L**ima
- **M**ike
- **N**ovember
- **O**scar
- **P**apa
- **Q**uebec
- **R**omeo
- **S**ierra
- **T**ango
- **U**niform
- **V**ictor
- **W**hiskey
- **X**ray
- **Y**ankee
- **Z**ulu

AMERICAN RADIO CLUB

International Morse code:

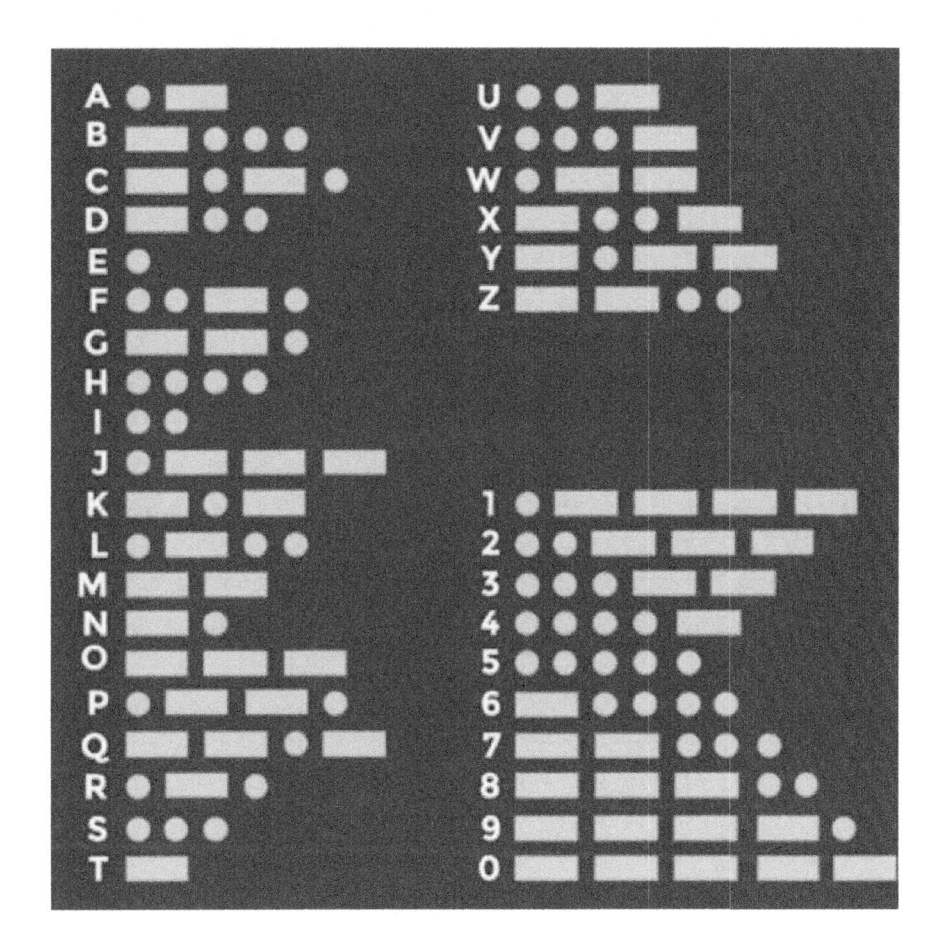

A	·−	U	··−
B	−···	V	···−
C	−·−·	W	·−−
D	−··	X	−··−
E	·	Y	−·−−
F	··−·	Z	−−··
G	−−·		
H	····		
I	··		
J	·−−−	1	·−−−−
K	−·−	2	··−−−
L	·−··	3	···−−
M	−−	4	····−
N	−·	5	·····
O	−−−	6	−····
P	·−−·	7	−−···
Q	−−·−	8	−−−··
R	·−·	9	−−−−·
S	···	0	−−−−−
T	−		

ITU region map:

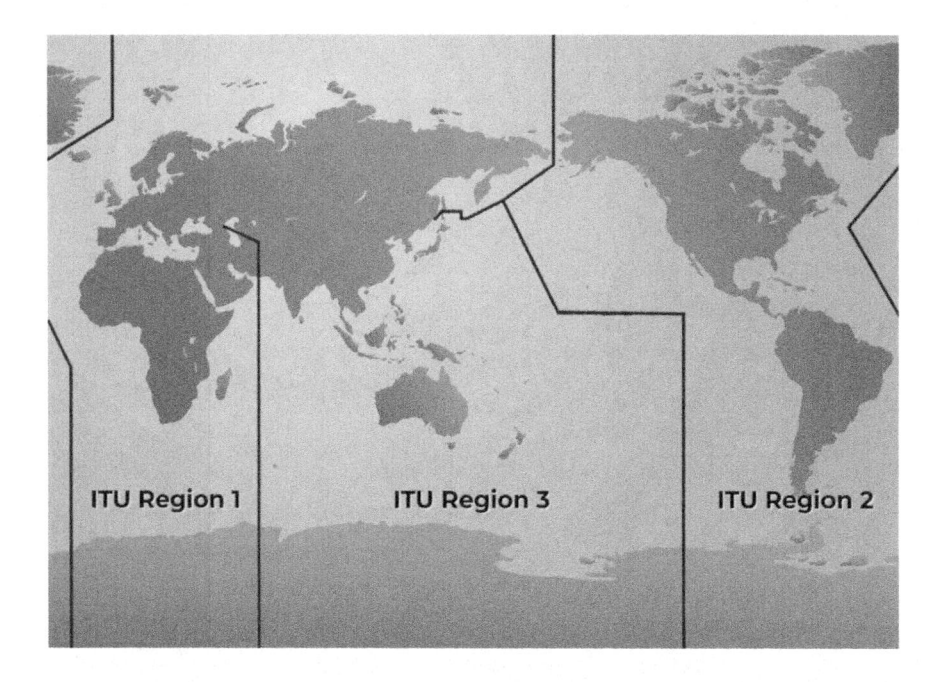

Q codes

Q-Code	Used as a Question	Used as a Statement
QRA	What is the name of your station?	My name is ...
QRB	How far approximately are you from my station?	The distance between our stations is about ... your nautical miles (or kilometers).
QRG	What is my exact frequency?	Your exact frequency is ... kHz (Or MHz).
QRK	What is the intelligibility of my signals?	The intelligibility of your signals is ... (scale of 1 to 5).
QRL	Are you busy?	I'm busy. Please do not interfere.
QRM	Are you bothered by noise?	I am disturbed by interference.
QRN	Are you bothered by noise of natural origin (storms, lightning)?	I am disturbed by natural origin noise.
QRO	Shall I increase transmitter power?	Increase the transmission power.
QRP	Shall I decrease transmitter power?	Decrease the transmission power.
QRQ	Shall I send faster?	Increase the transmission speed [... words per minute].
QRS	Shall I send more slowly?	Send more slowly [... words per minute].
QRT	Shall I stop transmissions?	Close (or I close) transmissions.
QRV	Are you ready?	I'm ready.
QRX	When will you call me again?	I'll get back at ... on ... kHz (or MHz).
QRZ	Who is calling me?	You are called by ... on ... kHz (or MHz).
QSA	What is the strength of my signals?	The strength of your signals is ... (Scale from 1 to 5).
QSB	Does my signal strength fade?	The strength of your signals varies.
QSK	Can you hear me? If so, can I interrupt you?	I hear you, speak up.
QSL	Can you receive?	Confirmed, received.
QSO	Can you communicate with ... directly or through support?	I can communicate with ... directly NOTE: It is also synonymous of direct communication or direct connection.
QRU	Have anything for me?	I have something for you
QSK	Can you hear me between your signals? If so may I break in?	I can hear you and you may break in
QST		Here is a bulletin for all amateur radio operators
QSY	Should I change my transmission to another frequency?	Change to transmission on another frequency
QTH	What is your location?	My location is ...

NOVICE CLASS
license frequency privileges

HF
200 watts PEP maximum

80 meters	3.525 - 3.600 MHz	CW
40 meters	7.025 - 7.125 MHz	CW
15 meters	21.025 - 21.200 MHz	CW
10 meters	28.000 - 28.300 MHz	CW, RTTY/data
	28.300 - 28.500 MHz	CW, phone

VHF
25 watts PEP maximum

1.25 meters 222.000 - 225.000 MHz   CW, MCW, phone, image, RTTY/data

UHF
5 watts PEP maximum

23 cm 1.270 - 1.295 GHz   CW, MCW, phone, image, RTTY/data

Notes : By FCC rules, amateur radio operators must operate only within the frequency ranges allowed by their class of license and they must ensure the bandwidth used keeps them within the stated ranges. The information contained in this graphic is accurate as of the date of publication to our knowledge, however, licensees need to keep up to date on their frequency privileges by monitoring FCC rules changes. You must have an amateur radio license from the Federal Communications Commission to transmit on any amateur radio frequencies. Transmitting without a license can result in heavy fines and/or imprisonment.

© AMERICAN RADIO CLUB

TECHNICIAN CLASS
license frequency privileges

HF
200 watts PEP maximum

80 meters	3.525	- 3.600 MHz	CW
40 meters	7.025	- 7.125 MHz	CW
15 meters	21.025	- 21.200 MHz	CW
10 meters	28.000	- 28.300 MHz	CW, RTTY/data
	28.300	- 28.500 MHz	CW, SSB

VHF
1,500 watts PEP maximum

6 meters	50.000	- 50.100 MHz	CW
	50.100	- 54.000 MHz	CW, digital, SSB, AM, FM, TV
2 meters	144.000	- 144.100 MHz	CW
	144.100	- 148.000 MHz	CW, digital, SSB, AM, FM, TV
1.25 meters	219.000	- 220.000 MHz	Point-to-point digital links - 50 watts PEP maximum; 100 kHz bandwidth
	222.000	- 225.000 MHz	CW, digital, SSB, AM, FM, TV

UHF
1,500 watts PEP maximum

70 cm	420.000	- 450.000 MHz	CW, digital, SSB, AM, FM, TV
	420.000	- 430.000 MHz	Not available for use north of Line A near Canada border
33 cm	902.000	- 928.000 MHz	CW, digital, SSB, AM, FM, TV
23 cm	1.240	- 1.300 GHz	CW, digital, SSB, AM, FM, TV
13 cm	2.300	- 2.310 GHz	CW, digital, SSB, AM, FM, TV
	2.390	- 2.450 GHz	CW, digital, SSB, AM, FM, TV

© AMERICAN RADIO CLUB

TECHNICIAN CLASS
license frequency privileges

SHF
1,500 watts PEP maximum

Band	Frequency	Modes
9 cm	3.300 - 3.500 GHz	CW, digital, SSB, AM, FM, TV
5 cm	5.650 - 5.925 GHz	CW, digital, SSB, AM, FM, TV
3 cm	10.000 - 10.500 GHz	CW, digital, SSB, AM, FM, TV
1.2 cm	24.000 - 24.250 GHz	CW, digital, SSB, AM, FM, TV

EHF
1,500 watts PEP maximum

Band	Frequency	Modes
6 mm	47.000 - 47.200 GHz	CW, digital, SSB, AM, FM, TV
4 mm	76.000 - 81.000 GHz	CW, digital, SSB, AM, FM, TV
2.5 mm	122.250 - 123.000 GHz	CW, digital, SSB, AM, FM, TV
2 mm	134.000 - 141.000 GHz	CW, digital, SSB, AM, FM, TV
1 mm	241.000 - 250.000 GHz	CW, digital, SSB, AM, FM, TV
All Above	275 GHz and above	CW, digital, SSB, AM, FM, TV

Notes: By FCC rules, amateur radio operators must operate only within the frequency ranges allowed by their class of license and they must ensure the bandwidth used keeps them within the stated ranges. The information contained in this graphic is accurate as of the date of publication to our knowledge, however, licensees need to keep up to date on their frequency privileges by monitoring FCC rules changes. You must have an amateur radio license from the Federal Communications Commission to transmit on any amateur radio frequencies. Transmitting without a license can result in heavy fines and/or imprisonment.

© AMERICAN RADIO CLUB

GENERAL CLASS
license frequency privileges

LF
1 watt EIRP maximum

2200 meters 135.7 - 137.8 kHz   CW, phone, image, RTTY/data

MF
5 watts EIRP maximum (except 1 watt EIRP in Alaska within 496 miles of Russia)

630 meters 472 - 479 kHz   CW, phone, image, RTTY/data

HF
1,500 watts PEP maximum (unless noted)

160 meters 1.800 - 2.000 MHz   CW, phone, image, RTTY/data

80 meters 3.525 - 3.600 MHz   CW, RTTY/data
              3.800 - 4.000 MHz   CW, phone, image

60 meters
100 watts ERP into antenna with 0 dBd gain

Five 2.8-kHz channels (channel center shown)

5.332 MHz	CW, phone, narrow digital modes per rules
5.348 MHz	CW, phone, narrow digital modes per rules
5.3585 MHz	CW, phone, narrow digital modes per rules
5.373 MHz	CW, phone, narrow digital modes per rules
5.405 MHz	CW, phone, narrow digital modes per rules

© AMERICAN RADIO CLUB

GENERAL CLASS
license frequency privileges

40 meters 7.025 - 7.125 MHz   CW, RTTY/data
              7.175 - 7.300 MHz   CW, phone, image

30 meters
200 watts ERP maximum
              10.100 - 10.150 MHz   CW, RTTY/data

20 meters 14.025 - 14.150 MHz   CW, RTTY/data
              14.225 - 14.350 MHz   CW, phone, image

17 meters 18.068 - 18.110 MHz   CW, RTTY/data
              18.110 - 18.168 MHz   CW, phone, image

15 meters 21.025 - 21.200 MHz   CW, RTTY/data
              21.275 - 21.450 MHz   CW, phone, image

12 meters 24.890 - 24.930 MHz   CW, RTTY/data
              24.930 - 24.990 MHz   CW, phone, image

10 meters 28.000 - 28.300 MHz   CW, RTTY/data
              28.300 - 29.700 MHz   CW, phone, image

VHF
1,500 watts PEP maximum

6 meters   50.000 - 50.100 MHz   CW
              50.100 - 54.000 MHz   CW, digital, SSB, AM, FM, TV

2 meters   144.000 - 144.100 MHz   CW
              144.100 - 148.000 MHz   CW, MCW, digital, SSB, AM, FM, TV

1.25 meters 219.000 - 220.000 MHz   Point-to-point digital links - 50 watts PEP maximum; 100 kHz bandwidth
              222.000 - 225.000 MHz   CW, MCW, digital, SSB, AM, FM, TV

© AMERICAN RADIO CLUB

GENERAL CLASS
license frequency privileges

UHF
1,500 watts PEP maximum

70 cm	420.000 - 450.000 MHz	CW, MCW, digital, SSB, AM, FM, TV
	420.000 - 430.000 MHz	Not available for use north of Line A near Canada border
33 cm	902.000 - 928.000 MHz	CW, MCW, digital, SSB, AM, FM, TV
23 cm	1.240 - 1.300 GHz	CW, MCW, digital, SSB, AM, FM, TV
13 cm	2.300 - 2.310 GHz	CW, digital, SSB, AM, FM, TV
	2.390 - 2.450 GHz	CW, digital, SSB, AM, FM, TV

SHF
1,500 watts PEP maximum

9 cm	3.300 - 3.500 GHz	CW, digital, SSB, AM, FM, TV
5 cm	5.650 - 5.925 GHz	CW, digital, SSB, AM, FM, TV
3 cm	10.000 - 10.500 GHz	CW, digital, SSB, AM, FM, TV
1.2 cm	24.000 - 24.250 GHz	CW, digital, SSB, AM, FM, TV

EHF
1,500 watts PEP maximum

6 mm	47.000 - 47.200 GHz	CW, digital, SSB, AM, FM, TV
4 mm	76.000 - 81.000 GHz	CW, digital, SSB, AM, FM, TV
2.5 mm	122.250 - 123.000 GHz	CW, digital, SSB, AM, FM, TV
2 mm	134.000 - 141.000 GHz	CW, digital, SSB, AM, FM, TV
1 mm	241.000 - 250.000 GHz	CW, digital, SSB, AM, FM, TV
All Above	275 GHz and above	CW, digital, SSB, AM, FM, TV

Notes : By FCC rules, amateur radio operators must operate only within the frequency ranges allowed by their class of license and they must ensure the bandwidth used keeps them within the stated ranges. The information contained in this graphic is accurate as of the date of publication to our knowledge, however, licensees need to keep up to date on their frequency privileges by monitoring FCC rules changes. You must have an amateur radio license from the Federal Communications Commission to transmit on any amateur radio frequencies. Transmitting without a license can result in heavy fines and/or imprisonment.

© AMERICAN RADIO CLUB

ADVANCED CLASS
license frequency privileges

LF
1 watt EIRP maximum

2200 meters 135.7 - 137.8 kHz   CW, phone, image, RTTY/data

MF
5 watts EIRP maximum (except 1 watt EIRP in Alaska within 496 miles of Russia)

630 meters 472 - 479 kHz   CW, phone, image, RTTY/data

HF
1,500 watts PEP maximum (unless noted)

160 meters 1.800 - 2.000 MHz   CW, phone, image, RTTY/data
80 meters   3.525 - 3.600 MHz   CW, RTTY/data
                3.700 - 4.000 MHz   CW, phone, image

60 meters
100 watts ERP into antenna with 0 dBd gain

Five 2.8-kHz channels (channel center shown)

5.332 MHz	CW, phone, narrow digital modes per rules
5.348 MHz	CW, phone, narrow digital modes per rules
5.3585 MHz	CW, phone, narrow digital modes per rules
5.373 MHz	CW, phone, narrow digital modes per rules
5.405 MHz	CW, phone, narrow digital modes per rules

© AMERICAN RADIO CLUB

ADVANCED CLASS
license frequency privileges

40 meters 7.025 - 7.125 MHz   CW, RTTY/data
7.125 - 7.300 MHz   CW, phone, image

30 meters
200 watts ERP maximum
10.100 - 10.150 MHz   CW, RTTY/data

20 meters 14.025 - 14.150 MHz   CW, RTTY/data
14.175 - 14.350 MHz   CW, phone, image

17 meters 18.068 - 18.110 MHz   CW, RTTY/data
18.110 - 18.168 MHz   CW, phone, image

15 meters 21.025 - 21.200 MHz   CW, RTTY/data
21.225 - 21.450 MHz   CW, phone, image

12 meters 24.890 - 24.930 MHz   CW, RTTY/data
24.930 - 24.990 MHz   CW, phone, image

10 meters 28.000 - 28.300 MHz   CW, RTTY/data
28.300 - 29.700 MHz   CW, phone, image

VHF
1,500 watts PEP maximum

6 meters 50.000 - 50.100 MHz   CW
50.100 - 54.000 MHz   CW, digital, SSB, AM, FM, TV

2 meters 144.000 - 144.100 MHz   CW
144.100 - 148.000 MHz   CW, MCW, digital, SSB, AM, FM, TV

1.25 meters 219.000 - 220.000 MHz   Point-to-point digital links - 50 watts PEP maximum; 100 kHz bandwidth
222.000 - 225.000 MHz   CW, MCW, digital, SSB, AM, FM, TV

© AMERICAN RADIO CLUB

ADVANCED CLASS
license frequency privileges

UHF
1,500 watts PEP maximum

70 cm	420.000 - 450.000 MHz	CW, MCW, digital, SSB, AM, FM, TV
	420.000 - 430.000 MHz	Not available for use north of Line A near Canada border
33 cm	902.000 - 928.000 MHz	CW, MCW, digital, SSB, AM, FM, TV
23 cm	1.240 - 1.300 GHz	CW, MCW, digital, SSB, AM, FM, TV
13 cm	2.300 - 2.310 GHz	CW, digital, SSB, AM, FM, TV
	2.390 - 2.450 GHz	CW, digital, SSB, AM, FM, TV

SHF
1,500 watts PEP maximum

9 cm	3.300 - 3.500 GHz	CW, digital, SSB, AM, FM, TV
5 cm	5.650 - 5.925 GHz	CW, digital, SSB, AM, FM, TV
3 cm	10.000 - 10.500 GHz	CW, digital, SSB, AM, FM, TV
1.2 cm	24.000 - 24.250 GHz	CW, digital, SSB, AM, FM, TV

EHF
1,500 watts PEP maximum

6 mm	47.000 - 47.200 GHz	CW, digital, SSB, AM, FM, TV
4 mm	76.000 - 81.000 GHz	CW, digital, SSB, AM, FM, TV
2.5 mm	122.250 - 123.000 GHz	CW, digital, SSB, AM, FM, TV
2 mm	134.000 - 141.000 GHz	CW, digital, SSB, AM, FM, TV
1 mm	241.000 - 250.000 GHz	CW, digital, SSB, AM, FM, TV
All Above	275 GHz and above	CW, digital, SSB, AM, FM, TV

Notes: By FCC rules, amateur radio operators must operate only within the frequency ranges allowed by their class of license and they must ensure the bandwidth used keeps them within the stated ranges. The information contained in this graphic is accurate as of the date of publication to our knowledge, however, licensees need to keep up to date on their frequency privileges by monitoring FCC rules changes. You must have an amateur radio license from the Federal Communications Commission to transmit on any amateur radio frequencies. Transmitting without a license can result in heavy fines and/or imprisonment.

© AMERICAN RADIO CLUB

EXTRA CLASS
license frequency privileges

LF
1 watt EIRP maximum

2200 meters   135.7 - 137.8 kHz   CW, phone, image, RTTY/data

MF
5 watts EIRP maximum (except 1 watt EIRP in Alaska within 496 miles of Russia)

630 meters   472 - 479 kHz   CW, phone, image, RTTY/data

HF
1,500 watts PEP maximum (unless noted)

160 meters   1.800 - 2.000 MHz   CW, phone, image, RTTY/data

80 meters   3.500 - 3.600 MHz   CW, RTTY/data
                 3.600 - 4.000 MHz   CW, phone, image

60 meters
100 watts ERP into antenna with 0 dBd gain

Five 2.8-kHz channels (channel center shown)

5.332 MHz	CW, phone, narrow digital modes per rules
5.348 MHz	CW, phone, narrow digital modes per rules
5.3585 MHz	CW, phone, narrow digital modes per rules
5.373 MHz	CW, phone, narrow digital modes per rules
5.405 MHz	CW, phone, narrow digital modes per rules

© AMERICAN RADIO CLUB

EXTRA CLASS
license frequency privileges

40 meters 7.000 - 7.125 MHz   CW, RTTY/data
              7.125 - 7.300 MHz   CW, phone, image

30 meters
200 watts ERP maximum
              10.100 - 10.150 MHz   CW, RTTY/data

20 meters 14.000 - 14.150 MHz   CW, RTTY/data
              14.150 - 14.350 MHz   CW, phone, image

17 meters 18.068 - 18.110 MHz   CW, RTTY/data
              18.110 - 18.168 MHz   CW, phone, image

15 meters 21.000 - 21.200 MHz   CW, RTTY/data
              21.200 - 21.450 MHz   CW, phone, image

12 meters 24.890 - 24.930 MHz   CW, RTTY/data
              24.930 - 24.990 MHz   CW, phone, image

10 meters 28.000 - 28.300 MHz   CW, RTTY/data
              28.300 - 29.700 MHz   CW, phone, image

VHF
1,500 watts PEP maximum

6 meters 50.000 - 50.100 MHz   CW
              50.100 - 54.000 MHz   CW, digital, SSB, AM, FM, TV

2 meters 144.000 - 144.100 MHz   CW
              144.100 - 148.000 MHz   CW, MCW, digital, SSB, AM, FM, TV

1.25 meters 219.000 - 220.000 MHz   Point-to-point digital links - 50 watts PEP maximum; 100 kHz bandwidth

              222.000 - 225.000 MHz   CW, MCW, digital, SSB, AM, FM, TV

© AMERICAN RADIO CLUB

EXTRA CLASS
license frequency privileges

UHF
1,500 watts PEP maximum

70 cm	420.000 - 450.000 MHz	CW, MCW, digital, SSB, AM, FM, TV
	420.000 - 430.000 MHz	Not available for use north of Line A near Canada border
33 cm	902.000 - 928.000 MHz	CW, MCW, digital, SSB, AM, FM, TV
23 cm	1.240 - 1.300 GHz	CW, MCW, digital, SSB, AM, FM, TV
13 cm	2.300 - 2.310 GHz	CW, digital, SSB, AM, FM, TV
	2.390 - 2.450 GHz	CW, digital, SSB, AM, FM, TV

SHF
1,500 watts PEP maximum

9 cm	3.300 - 3.500 GHz	CW, digital, SSB, AM, FM, TV
5 cm	5.650 - 5.925 GHz	CW, digital, SSB, AM, FM, TV
3 cm	10.000 - 10.500 GHz	CW, digital, SSB, AM, FM, TV
1.2 cm	24.000 - 24.250 GHz	CW, digital, SSB, AM, FM, TV

EHF
1,500 watts PEP maximum

6 mm	47.000 - 47.200 GHz	CW, digital, SSB, AM, FM, TV
4 mm	76.000 - 81.000 GHz	CW, digital, SSB, AM, FM, TV
2.5 mm	122.250 - 123.000 GHz	CW, digital, SSB, AM, FM, TV
2 mm	134.000 - 141.000 GHz	CW, digital, SSB, AM, FM, TV
1 mm	241.000 - 250.000 GHz	CW, digital, SSB, AM, FM, TV
All Above	275 GHz and above	CW, digital, SSB, AM, FM, TV

Notes: By FCC rules, amateur radio operators must operate only within the frequency ranges allowed by their class of license and they must ensure the bandwidth used keeps them within the stated ranges. The information contained in this graphic is accurate as of the date of publication to our knowledge, however, licensees need to keep up to date on their frequency privileges by monitoring FCC rules changes. You must have an amateur radio license from the Federal Communications Commission to transmit on any amateur radio frequencies. Transmitting without a license can result in heavy fines and/or imprisonment.

© **AMERICAN RADIO CLUB**

Get more involved in amateur radio by upgrading your license

American Radio Club offers in-depth, fun and affordable courses for those who want to take it to the next level.

Whether you're a Novice or Technician who wants access to more frequencies or power, or an experienced General or Advanced licensee who wants to take that final leap into the Amateur Extra class so you finally can have full amateur spectrum privileges, Ham Radio Prep offers something for everyone.

Tens of thousands of American Radio Club students have succeeded in getting licensed easily with these courses:

Technician: AmericanRadioClub.com/technician-license-course

General: AmericanRadioClub.com/general-license-course

Amateur Extra: AmericanRadioClub.com/amateur-extra-license

Made in the USA
Monee, IL
06 June 2022